Electrons

Or the Nature and Properties
of Negative Electricity

OLIVER LODGE

CAMBRIDGE
UNIVERSITY PRESS

CAMBRIDGE UNIVERSITY PRESS

Cambridge, New York, Melbourne, Madrid, Cape Town,
Singapore, São Paolo, Delhi, Mexico City

Published in the United States of America by Cambridge University Press, New York

www.cambridge.org
Information on this title: www.cambridge.org/9781108052146

© in this compilation Cambridge University Press 2012

This edition first published 1906
This digitally printed version 2012

ISBN 978-1-108-05214-6 Paperback

CAMBRIDGE LIBRARY COLLECTION

Books of enduring scholarly value

Physical Sciences

From ancient times, humans have tried to understand the workings of the world around them. The roots of modern physical science go back to the very earliest mechanical devices such as levers and rollers, the mixing of paints and dyes, and the importance of the heavenly bodies in early religious observance and navigation. The physical sciences as we know them today began to emerge as independent academic subjects during the early modern period, in the work of Newton and other 'natural philosophers', and numerous sub-disciplines developed during the centuries that followed. This part of the Cambridge Library Collection is devoted to landmark publications in this area which will be of interest to historians of science concerned with individual scientists, particular discoveries, and advances in scientific method, or with the establishment and development of scientific institutions around the world.

Electrons

The discovery in 1897 of the electron, the first subatomic particle, led to rapid advances in our knowledge of atomic structure, the solid state, radioactivity and chemistry. It also raised major questions. Was the electron point-like or did it have structure? Was there a positive electron? What did the positive part of the atom look like? Did a hydrogen atom have one electron or a thousand? Published in 1906, this expository account by leading physicist Sir Oliver Lodge (1851–1940) examines the spectacular phenomena of cathode rays in evacuated tubes, the fixed units of charge observed in electrolysis, and the puzzling regularities in atomic spectra. Lodge knew most of the pioneers in the field, and his enthusiastic descriptions of their work and clear analyses of the problems as well as successes paint a vivid picture of the excitement of cutting-edge research and the scientific process in action.

Cambridge University Press has long been a pioneer in the reissuing of out-of-print titles from its own backlist, producing digital reprints of books that are still sought after by scholars and students but could not be reprinted economically using traditional technology. The Cambridge Library Collection extends this activity to a wider range of books which are still of importance to researchers and professionals, either for the source material they contain, or as landmarks in the history of their academic discipline.

Drawing from the world-renowned collections in the Cambridge University Library and other partner libraries, and guided by the advice of experts in each subject area, Cambridge University Press is using state-of-the-art scanning machines in its own Printing House to capture the content of each book selected for inclusion. The files are processed to give a consistently clear, crisp image, and the books finished to the high quality standard for which the Press is recognised around the world. The latest print-on-demand technology ensures that the books will remain available indefinitely, and that orders for single or multiple copies can quickly be supplied.

The Cambridge Library Collection brings back to life books of enduring scholarly value (including out-of-copyright works originally issued by other publishers) across a wide range of disciplines in the humanities and social sciences and in science and technology.

ELECTRONS, OR THE NATURE AND
PROPERTIES OF NEGATIVE ELECTRICITY

GEORGE BELL AND SONS
LONDON: PORTUGAL ST., LINCOLN'S INN
CAMBRIDGE: DEIGHTON, BELL AND CO.
NEW YORK: THE MACMILLAN COMPANY
BOMBAY: A. H. WHEELER AND CO.

ELECTRONS

OR

THE NATURE AND PROPERTIES
OF NEGATIVE ELECTRICITY

BY

SIR OLIVER LODGE, F.R.S.

D.Sc. Lond., Hon. D.Sc. Oxon. et Vict., LL.D. St. Andrews, Glasgow, and Aberdeen

VICE-PRESIDENT OF THE INSTITUTION OF ELECTRICAL ENGINEERS
RUMFORD MEDALLIST OF THE ROYAL SOCIETY
EX-PRESIDENT OF THE PHYSICAL SOCIETY OF LONDON
LATE PROFESSOR OF PHYSICS IN THE UNIVERSITY COLLEGE OF LIVERPOOL
HONORARY MEMBER OF THE AMERICAN PHILOSOPHICAL SOCIETY OF PHILADELPHIA, OF THE
BATAVIAN SOCIETY OF ROTTERDAM, AND OF THE ACADEMY OF SCIENCES OF BOLOGNA
PRINCIPAL OF THE UNIVERSITY OF BIRMINGHAM

LONDON
GEORGE BELL AND SONS
1906

GLASGOW : PRINTED AT THE UNIVERSITY PRESS
BY ROBERT MACLEHOSE AND CO. LTD.

TO THE

CAVENDISH PROFESSORS OF PHYSICS

IN THE UNIVERSITY OF CAMBRIDGE, AND ESPECIALLY

TO THE PRESENT HOLDER OF THE CHAIR, THIS SMALL BOOK IS

DEDICATED WITH PROFOUND ADMIRATION

BY THE AUTHOR

PREFACE

In 1902 I was asked by the President of the Institution of Electrical Engineers to give to that body a discourse on recent progress towards knowledge of the nature of Electricity, especially concerning its discontinuous or atomic structure. This discourse, greatly extended, appeared in Vol. 32 of the Journal of the Institution, and constitutes the nucleus of the present book.

Many additions have now been made, and some of the difficulties recently promulgated concerning the possibility of an electric theory of matter are touched upon. They are of date too recent to have been mentioned even in my "Romanes Lecture" before the University of Oxford, published under the title *Modern Views of Matter* by the Clarendon Press.

The most important addition is a more detailed account of the proof of the purely electrical nature of the mass or inertia of an electron : an investigation generally associated on the experimental side with the name of Kaufmann, but of course based on the work of many predecessors and contemporaries. A proof that the atom of matter is essentially composed of such electrons, and that its mass too is of purely electromagnetic nature, is lacking : the electromagnetic theory of Matter, unlike the electromagnetic theory of Light, must be regarded for the present as no better than a working hypothesis. It is a hypothesis of stimulating character, and of great probability, but its truth is still an open question that is probably not going to be speedily closed.

I am indebted to Professor Larmor for information about some recent theoretical work, and for the substance of Appendix M; I have also to thank Mr. Gwilym Owen, of the University of Liverpool, for assistance in the revision of the proof.

As an introduction to an allied subject, the book called *Becquerel Rays*, by the Hon. R. J. Strutt, is to be recommended; and the standard treatise of Professor Rutherford on *Radioactivity* is well known. I have avoided dealing at length with the topics so conveniently to be found in these writings. I have also barely touched on the large subject of 'ionisation': it was difficult to do so without overloading the principles with detail, a knowledge of which is nevertheless necessary for investigators. The treatise of Prof. J. J. Thomson, *The Discharge of Electricity through Gases*, contains a mass of information and original work highly valued by physicists.

The present book is intended throughout for students of general physics, and in places for specialists, but most of it may be taken as an exposition of a subject of inevitable interest to all educated men.

OLIVER LODGE.

THE UNIVERSITY OF BIRMINGHAM,
July, 1906.

CONTENTS

CONTENTS xi

INTRODUCTION

In Maxwell's *Electricity* published in 1873, section 57, the following sentence occurs in connection with the discharge of electricity through gases, especially through rarefied gases :

> "These and many other phenomena of electrical discharge are exceedingly important, and, when they are better understood, they will probably throw great light on the nature of electricity as well as on the nature of gases and of the medium pervading space."

This prediction has been amply justified by the progress of science, and, no doubt, still further possibilities of advance lie in the same direction. The study of conduction through liquids, first, and the study of conduction through gases, next, combined with a study of the processes involved in radiation, have resulted in an immense addition to our knowledge of late years, and have opened a new chapter, indeed a new volume, of Physics.

The net result has been to concentrate attention upon the phenomena of electric charge, and greatly

to enhance the importance of a study of electrostatics. Not long ago our brilliant and lamented
friend, G. F. FitzGerald, used chaffingly to speak
of electrostatics as "one of the most beautiful and
useless adaptations of nature"; and it was becoming
the custom with teachers, who felt that they must
attend exclusively to the practically useful and not
waste their students' time on decoration and super-
fluities, almost to ignore, or at any rate to scamper
through, the domain of electrostatics, and to begin
the study of electricity with the phenomena of
current, especially with the connection between
electricity and magnetism.

And certainly from the severely practical point
of view, as well as from many other aspects, this
part of electrical science remains the most important; but to him who would not only design
dynamos and large-scale machinery, to him who, in
addition to the training and aptitude of the
engineer, possesses something of the interests, the
instinct, and the insight, of a man of science,—to
such a one the nature and properties of an electric
charge, at rest and in motion, constitute a fascinat-
ing study; for there lies the key to the inner
meaning of all the occurrences with which his active
life is so intimately concerned—there lies the
proximate solution of problems which have excited
the attention and taxed the ingenuity of philosophers and physicists and chemists for more than
a century. Indeed it turns out that subjects

broader and more fundamental than those known as 'electric' are indirectly involved; and we are now beginning to have some hope of obtaining unexpected answers to riddles—such as those concerning the fundamental properties of matter—which have proposed themselves for solution throughout the history of civilisation. Problems of this kind have aroused interest and attention ever since men began to escape from the struggle for bare existence—that most immediately practical of all occupations—and felt free to devote themselves, some to art, some to literature, some to the accumulation of superfluous wealth, and some to the gratuitous pursuit of philosophical speculation, exact experiment, and pure theory. To this comparatively leisured group I now address myself.

MEANING OF TERMS AND SYMBOLS AS USED IN THIS BOOK.

Electron = the unit electric charge, or atom of negative electricity.

e = the amount of this charge, whether positive or negative; about 3×10^{-10} electrostatic unit.

E = an electromotive force, or the strength of an electric field.

H = the strength of a magnetic field.

m = mass or inertia, especially the mass of an electron.

a = usually, the linear dimension of an electron (about the hundred-thousandth of b).

b = the linear dimension of an atom of matter (the ten-millionth of a millimetre).

ion = an atom of matter with an unbalanced electric charge, either negative or positive, attached to it : the cause of chemical affinity.

κ = Faraday's dielectric constant, or the specific inductive capacity of free ether.

μ = the magnetic permeability of free ether.

[These two are the great ethereal constants, whose value and nature are not yet known. Only their product is known.]

v = the velocity of light *in vacuo* = $1/\sqrt{(\mu\kappa)}$
= 3×10^{10} cm. per sec.

u = the velocity of a particle.

CHAPTER I.

PROPERTIES OF AN ELECTRIC CHARGE.

FIRST I must lay a basis of pure theory : we must consider the properties of the ancient and long known phenomenon called an electrified body.

Two substances placed in intimate contact and separated are in general united more or less permanently by lines of force, the region between them being in a state of tension along the lines and of pressure at right angles. These lines have direction and 'sense'—their two ends are not alike : they begin at one body and end at another, they map out a field of electrostatic force, and their terminations on one or other of the bodies constitute what we call an electric charge. Electric charges are of two kinds, positive and negative, the former corresponding to the beginning of the lines, the latter to their ends. To one class of bodies, called insulators, the lines appear rigidly attached : the charges cannot be displaced nor transferred elsewhere without violence ; whereas in another class they slip easily along, and are transferred from one such conducting body to another in contact with it, with great ease.

A tension in the lines tends to bring the ends together as near as possible, while laterally the lines tend to drive each other apart : this image sufficing to represent all that is observed as electrical attractions and repulsions.

The field of force mapped by the lines can exist in vacuo perfectly well, but the lines never terminate in vacuo; the charges are always carried by matter, or by something equivalent thereto. In empty space it is probable that the only way of destroying such a field of force is to allow the two bodies possessing the charges to approach each other, and thus shorten up the lines to nothing; though, even so, it is not probable that the charges are destroyed, but only placed so close together that they have no external effect at any moderate distance. When matter is present, however, it may be able to assist this collapse of the lines in various ways, giving rise to the various phenomena of conduction and of disruptive discharge.

If one of the two oppositely charged bodies is sent away to a considerable distance, while the other is isolated and regarded alone, the lines of this latter start out in all directions in nearly straight lines, giving rise to the simple notion of a single charged body. There must however be a complementary charge,—the other ends of the lines must be somewhere; though they may be so far away as to be spoken of as, for all practical purposes, at infinity.

Parenthetical remark of general application. People often feel hesitation about the treatment of things as at infinity, as if it introduced a conception of some difficulty; but they should realise that this mode of expression is always employed as a simplification, whenever it happens that for present purposes the said things can be ignored. If their existence requires attention, it must be recognised that they are really at some finite distance, and their location must be specified; but such specification complicates both ideas and equations. Whenever attention to them is unnecessary, or their location immaterial, this specification is avoided by treating them practically as if they were at infinity, that is by ignoring them. Every now and then this policy of ignoration must be suspended, but for a multiplicity of purposes it serves.

Charge in uniform motion.

Now consider how far this field of force belongs to the body, and how far it belongs to space, that is to the ether surrounding the body. The body is the nucleus whence the lines radiate, but the lines themselves, the state of tension and other properties which they represent and map out, do not belong to the body at all; at each point of space there is a peculiar etherial condition called an electric potential, and this potential represents something occurring in the ether and in the ether alone, though it is originated and maintained by the body.

Picture in the mind's eye such a charged body, say a charged sphere, and let it change its position; how are we to regard the effect of the displacement on its field of force? Few things in physics are more certain than this, that when a body moves along, the ether in its neighbourhood is not dragged with it, as if it were in the slightest degree viscous.* The ether, in fact, as a whole—*i.e.* when unmodified or in its normal condition—is stationary: it is susceptible to strain, but not to motion; it is the receptacle of potential, not of locomotive kinetic energy. The only generated motion to which it is possibly susceptible is of what is called 'irrotational' character—in other words it behaves as a perfect fluid. It may possess

---

* Lodge, *Phil. Trans.* 1893, p. 727, and especially 1897, p. 149.

other rotational; or vortex-like motion, but, if so, it is indestructible and unproducible by any known means, and has not yet been discovered.

The effect of the motion of the body, then, is to relieve the strain of the ether at one place and to generate it at another; the state of strain travels *with* the body, but *through* the ether.

Regarding the matter from the point of view of the ether, we might say that the field of force is constantly being destroyed and regenerated as the body moves. Regarding it from the point of view of the moving body, we should say that it carries its field with it.

The question now arises—and it is far from being an easy question—what sort of occurrences go on in the ether when this decay and regeneration of an electrostatic field is occurring, or when a field of force is moving through it? Can it adapt itself instantly to the new conditions, or does it require time? This matter has been studied, closely and exhaustively, by Mr. Oliver Heaviside.

Fix the eye upon a point a mile distant from the body; does the information about the motion of the body reach that point instantaneously, so that all the lines of force move like absolutely rigid spokes, every part simultaneously? If so, how is the communication carried on, so that the distant parts of the medium can be thus instantaneously affected? Or does the disturbance only arrive at the distant point after the lapse of a small but appreciable time; in other words, has there to be an adjustment to the new conditions—an adjustment which reaches the nearest parts first and the further parts later; and if so, what additional phenomena can be observed during the unsettled period?

The answer is that during the motion of the charged body, and even after the cessation of its motion,—until the disturbance has had time to die away and everything to settle down into static condition again,—the phenomena of *magnetism* make their appearance : a new set of lines of force quite different from the electrostatic lines (although they, too, exhibit a tension along them and a pressure at right angles) come into temporary being. These do not—like the electric ones—originate at one place and terminate at another : they are always and necessarily closed curves or rings, and in the present simple case they are circles all centred upon the path of motion of the charged body. At any point of space there are now three directions to consider : (1) there is the original direction of the electrostatic field—the original electric line of force ; (2) there is the direction of the motion—that is, a direction parallel to the movement of the charged sphere ; and (3) there is the direction at right angles to these two ; this last being the direction of the magnetic lines of force—the direction of the magnetic field.

I spoke of the magnetic field as temporary, but that is on the assumption that the charged body is merely displaced—merely shifted from one position to another ; if it is not stopped, but keeps on moving, then the magnetic lines continue as long as the motion lasts. The strength of the magnetic field, at any point with polar coordinates r, θ, is—

$$H = \frac{eu}{r^2} \sin \theta.$$

If we are asked whether such a magnetic field is weak or not, I have to reply that that depends entirely on how strong the charge is and how quickly it is moving. There is, in my opinion, no other kind of magnetic field possible ; and so if ever we come across a magnetic field which we feel entitled to consider "strong," we must conclude that it is associated with the motion of a very considerable charge, at a velocity we may properly style great. But certainly it is true that for any ordinary charged sphere, moving at any ordinary pace,—even supposing that it is a cannon-ball shot from the mouth of a gun—the concentric circular magnetic field surrounding its trajectory is decidedly feeble. Feeble or not, it is there, and to its existence we must trace all the magnetic phenomena of the electric current.

For just as there is no electrostatic field save that extending from one charged body to another, so there is no electric current except the motion of such a charged body, and no magnetic field except that which surrounds the path of this motion.

The locomotion of an electric charge *is* an electric current, and the magnetic field surrounding that current is believed to be the only kind of magnetic field in existence. If any other variety is possible, the burden of proof rests on those who make the positive assertion.

Transmission of Energy.

While the charge is stationary everything is steady, and we have an electric field only.

While the charge is moving at constant speed the current is steady, and we have a steady magnetic field superposed upon a steadily moving electric field; there is likewise a certain conveyance of energy in the direction of the motion.

This is a special case of the general theorem known as Poynting's : viz. that wherever an electric and a magnetic field are superposed there cannot be static equilibrium at that place, energy must flow through the medium; and the rate of transfer of energy—the amount conveyed per second through unit area—is equal to $1/4\pi$ times the vector product of the intensity of the two fields : that is to say, it is measured by the area of the parallelogram bounded by lines representing the two fields in magnitude and direction; a quantity commonly expressed as

$$V(EH), \text{ or } [EH], \text{ or as } EH \sin \theta,$$

where θ is the angle between E and H.

The direction of propagation of energy is normal to that same area, and its 'sense' or sign depends upon the sense of the two fields. If both were reversed, the sense of the transmission of energy would continue unchanged : and its amount remains constant so long as the fields are constant, that is so long as the current is steady. Another way of expressing the facts is to say that the space in which two fields are superposed is full of momentum; and that the moment of momentum appropriate to a pole m and a charge e is simply em.

Accelerated Charge.

One more statement :

So far we have dealt with the case of steady rest or steady motion; but what about the intermediate stages, the stages of starting and stopping? What is the condition of things after the charge has begun to move but before it has attained a constant speed, and again when the brake is applied and the speed is decreasing, or when the direction of motion is changing? What phenomena are observable during the epoch of acceleration or retardation of speed or curvature of path? Something more than simple electrostatics and simple magnetism is then observed.

For whenever a conductor is moved across a magnetic field it is well known that an electromotive force acts in that conductor, of magnitude equal to the rate at which magnetic lines of force are being cut; or in symbols

$$E = -dN/dt,$$

which is the fundamental 'dynamo' equation. This is called the phenomenon of magneto-electric induction; it is the induced E.M.F. discovered by Faraday, and it necessarily occurs whenever magnetism and relative motion are superposed.

It is quite independent of the conductivity of the conductor however, and would have the same value if the motion took place in an insulator, though of course it could not then produce the same effect as regards conduction-currents.

The effect of a conductor is to integrate, or add up, the E.M.F.'s generated in each element all along its length, and thus to display the effect in an obvious manner: especially when the conductor is made very long and is compactly coiled (as in an armature). The definition of electromotive force between two points A and B, or round any closed contour, is—the line-integral of electric field from A to B, or round the same contour. In the unclosed-path case it is measured by the difference of electric potential between A and B.

One of the easiest and most ordinary ways of superposing motion and magnetism, is to allow or cause a magnetic field to vary in strength (as in a Ruhmkorff coil); for then the lines of force move broadside on, expanding or contracting as the case may be, and thus at once we get the phenomenon of *induction*—the generation of an induced E.M.F., of value at any point equal to the rate of change of the lines of magnetic force there. This is what happens whenever an electric charge is accelerated; for then its magnetic field—which, as we have seen, depends upon the velocity—necessarily changes in strength, and so an E.M.F. is induced. There being no conductor, this E.M.F. will propel no current, but it will represent an electric force which was not there before, and the new force will be in a new direction; the direction of an induced electric force is perpendicular to the direction in which the growing magnetic lines are moving, which in the present case is outwards from the charge. Consequently the new or induced E.M.F. points in the direction of motion, though in the sense opposed to any change in it; and the effect of the superposition of this new E.M.F. upon the already existing magnetic field is to cause a certain small transmission of energy in a radial direction out and away from the accelerated charge. Some energy therefore flashes away with the speed of light; and although in ordinary cases it may be an exceedingly small amount which is thus radiated into space, yet it is the only mode of generating radiation with which we are acquainted.

It is from an electric charge during its epochs of acceleration or retardation that we get the phenomenon called *radiation*; it is this and this alone which excites ethereal waves, and gives us the different varieties of *light*.

The energy radiated per second has been shown by Larmor to be

$$\frac{2\mu e^2 \dot{u}^2}{3v},$$

where v is the speed of light and $\dot{u}$ is the acceleration of the charge e.

After this manner, though of course by means of a very extensive development of these fundamental ideas, are all the phenomena of electricity and magnetism and optics summarised, and, so to speak, accounted for.

NOTE.

Let it be said here, once for all, that in every case one or other of the two etherial constants, μ and κ, should be exhibited explicitly wherever it rightly occurs. If this be not done the dimensions of most expressions are necessarily wrong; and words have to be added about whether the units intended are c.g.s. or some other system, and likewise whether they are electrostatic or electromagnetic. This latter double system of measurement has served its turn, and still serves it; but intrinsically it is confusing, and has been only rendered necessary because we do not yet know the values, or even for certain the nature, of μ and κ. It is to be hoped that no third system—devised as an attempted escape from confusion, but really an intensification of fog—will ever be successfully attempted; though there are threats in that direction, owing to lack of clear thinking.

The explicit retention of the constants keeps everything clear and easy. For instance, the expression just above quoted is essentially what it purports to be, an energy divided by a time

$$\left[\frac{FL}{T}\right]$$

and is therefore true as it stands in every complete system of units whatever. So also the expression quoted near the beginning of the next chapter is essentially what *it* purports to be, namely, a mass or inertia; and the same may be said of all other expressions in this book.

The ordinary convention for numerical specification is,—for electromagnetic c.g.s. units, to consider $\mu = 1$, for electrostatic units, to consider $\kappa = 1$; and this convention must hold until we learn the real facts, by future discovery : for which discovery continual familiarity with the unknown constants undoubtedly serves to pave the way.

CHAPTER II.

ELECTRIC INERTIA.

Whatever a charge may be, and whatever the physical constitution of the ether, it must be able to maintain electric lines and magnetic lines separately, and to transmit energy wherever both sets of lines coexist and cross each other.

An accelerated charge is equivalent to a changing current, for dC/dt may be written d^2e/dt^2. Whenever a current changes it is well known that an E.M.F. of self-induction is set up, equal to $L\,dC/dt$; and this electrical equation $E = L\,\ddot{e}$ corresponds to the mechanical equation $F = m\,\ddot{x}$,—Newton's second law.

Considered from the point of view of a current as constituted by a moving charge, this self-induced or *reaction* E.M.F. corresponds or is analogous to a mass-acceleration. And the electrical acceleration is opposed by the E.M.F., just as the acceleration of matter is opposed by its mechanical inertia. The coefficient of the electric acceleration—commonly denoted by L—represents therefore an inertia term, and is properly called 'electric inertia.'

By Lenz's law the effect of induction is always to oppose the cause which is producing it. In the present case the 'cause' is the acceleration or retardation of the moving charge; and so, in each case, this is opposed by the reaction of the magnetic lines generated by it.

Motion is opposed while it is increasing in speed, and it is assisted while it is decreasing in speed—an effect precisely analogous to ordinary mechanical inertia;—and therefore force is necessary, and work must be done, either to start or to stop the motion of a charged body. An *extra* force, that is, by reason of its charge. Whatever ordinary inertia the body may have, considered as a piece of matter, it has a trifle more by reason of its being charged with electricity—no matter what the sign of its charge may be.

The value of this imitation or electrical inertia, for the case of a charged sphere of radius a, was calculated by J. J. Thomson in 1881, and is

$$\frac{2\mu e^2}{3a}.$$

Electrical Inertia or Mass, continued.

Since this is very important, I repeat:—

Just as a changing magnetic field affects an electrostatic charge,—that is to say generates a feeble field of electric force, into the intensity of which the velocity of light enters squared in the denominator (see Lodge, *Phil. Mag.*, June 1889, p. 472),—so it is with a changing electric field, it generates a magnetic field proportional to its velocity of change. And if it is being accelerated, the magnetic field itself varies, and in that case generates an E.M.F. which reacts upon the accelerated moving charge, —always in such a way as to oppose its motion—by what is called Lenz's law, or simply by the law of conservation of energy: for if it assisted the motion, the action and reaction would go on intensifying themselves, until any amount of violence was reached.

The magnetic lines generated by a rising current, that is by a positively accelerated charged body, react back upon the motion which produced them in such a way as to oppose it;—to oppose it actively or elastically, not passively or sluggishly as by friction. The reaction ceases the instant the motion becomes steady : it is not analogous to friction therefore, but to inertia; it is the coefficient of an acceleration term.

The magnetic lines generated by a falling current, that is by a negatively accelerated or retarded charged body, react oppositely, and tend to continue the motion : thus here also we have a term corresponding to inertia. And the charged body may be said to have extra momentum, by reason of its charge, while it is moving. The value of the momentum is proportional to the velocity, so long as the velocity is not excessively great; and accordingly the inertia term is constant, *i.e.* independent of speed, under the same restriction. It may therefore be considered to be in existence even when the charge is stationary, and thus it simulates exactly the familiar mechanical inertia of a lump of ordinary matter.

In Appendix A, is given the simplest form of the quantitative relations here indicated, and the inertia due to an electric charge is there calculated. It is to be understood that whatever inertia a material sphere may possess, considered as matter, it will possess more when it is charged with electricity, and this no matter whether the charge be positive or negative. The amount of extra or

electrical inertia is proportional to the electrostatic
energy of the charge: that is to say, it is proportional to the charge and its potential conjointly.
Call the charge e, and the radius of the sphere a,
the potential will be $e/\kappa a$ (κ being Faraday's
dialectric constant); and the appropriate inertia is

$$m = \frac{2}{3v^2}e \cdot \frac{e}{\kappa a},$$

where v is the velocity of light. (See Note at end
of chap. i.)

Another way of putting it is to say that if a
real mass of this amount were moving with the
speed of light, its kinetic energy would be half as
great again as the potential energy of the electric
charge thus reckoned as mechanically equivalent
to it;

for $\quad \dfrac{3}{4}mv^2 = \dfrac{1}{2}e \cdot \dfrac{e}{\kappa a} = \dfrac{1}{2}$ charge × potential

$$= potential\ energy.$$

Now any appreciable quantity of matter, even a
milligramme, moving with the speed of light, must
have a prodigious amount of energy; for, on the
ordinary assumption that mass is quite constant,
the energy of one milligramme rushing along with
the light-speed would amount to no less than
fifteen million foot-tons. Or as Sir William Crookes
has expressed it: a gramme, or fifteen grains, of
matter, moving with the speed of light, would have
energy enough to lift the British Navy to the top
of Ben Nevis.

Consequently the inertia of any ordinary quantity
of electric charge must be exceedingly minute.
Notwithstanding this, it is quite doubtful whether

or not there really exists any other kind of inertia. The question whether there does or not is at present, strictly speaking, an open one.

Effect of Concentration.

The only way of conferring upon a given electric charge any appreciable mass, is to make its potential exceedingly high, that is to concentrate it on a very small sphere.

A coulomb at the potential of a volt has an electrostatic energy of half a Joule, that is $\frac{1}{2} \times 10^7$ ergs.

The mass equivalent to this would be

$$\frac{2}{3}\frac{10^7}{9 \times 10^{20}} = \frac{2}{27} \times 10^{-13} \text{ gramme} = 10^{-8} \text{ milligramme.}$$

Raise the potential to a million volts, and the mass-equivalent to a coulomb at that potential would be the hundredth part of a milligramme; still barely appreciable therefore.

The charge on an atom as observed in electrolysis is known to be 10^{-10} electrostatic units.* If this were distributed uniformly on a sphere the nominal size of an atom, viz., one 10^{-8} centimetre in radius, its potential would be one hundredth of an electrostatic unit, or about 3 volts. The energy of such a charge would be 10^{-12} erg, and the inertia of a body which would possess this energy if moving at the speed of light would be 10^{-33} gramme; which would therefore be its electrical inertia or extra mass.

But this is incomparably smaller than the mass of a hydrogen atom, which is approximately 10^{-25} gramme. Consequently the ionic charge distributed uniformly over an atom would add no appreciable fraction to its apparent mass.

* More exactly, according to Cambridge measurements, $3\cdot3 \times 10^{-10}$.

If, however, the atomic charge were concentrated into a sphere of dimension 10^{-13} centimetre, its potential would be 1000 electrostatic units, or 300,000 volts; its energy would then be 10^{-7} erg, and its inertia 10^{-28} gramme, or about $\frac{1}{1000}$ of the mass of a hydrogen atom.

Summary.

All this is a preliminary statement of undeniable fact : that is to say of fact which follows from the received and established theory of Electricity, whether such things as electrons have ever been found to exist or not.

All that we have stated is true of an ordinary charge on any ordinary sphere which can be made to move by mechanical force applied to it.

It gives us the phenomena
of electrostatics when at rest,
of magnetism when in motion,
of radiation when its motion is altered ;
and it incidentally, by reason of the known laws of electromagnetic induction, exhibits a kind of imitation inertia, and in that way simulates the possession of the most fundamental property of matter.

I will add a few more closely connected assertions, for later application :

Apply a sufficiently violent E.M.F. to a charged sphere, and the charge may be wrenched off it.

Insert an obstacle in the path of a violently moving charged sphere, so as to stop it mechanically with *sufficient* suddenness, and again it is possible for the charge, or something like it, to be jerked off it and passed on. But to do this the speed of the material sphere, as well as the suddenness of stoppage, must be excessive. Usually the charge is merely thrown into oscillation, when the sphere is suddenly stopped; and it then emits a solitary wave or spherical shell of thickness equal to the diameter of the sphere : or greater than that diameter by the amount the sphere has moved during its retardation. When the acceleration is moderate, however, the radiation is less energetic and also less intense : less energetic because its power depends on the square of the acceleration, less intense because it is spread over a thicker ethereal shell. Röntgen rays are perceptible only when the speed was great and the stoppage so sudden that the wave or pulse-shell is strong and thin (see chap. viii.). The thinner the pulses or wave shells the more penetrating they are. If thin enough they could traverse matter without affecting it or being affected by it.

Historical Remarks.

The doctrine of the behaviour of a charged sphere in motion, and the calculation of the value of the quasi inertia of an electric charge, was begun by Professor J. J. Thomson in an epoch-making paper published in the *Philosophical Magazine* for April, 1881—one of the most remarkable physical memoirs of our time.

The stimulus to this investigation was supplied by those brilliant experiments of Crookes, published in the *Philosophical Transactions* for 1879, which were preceded by observations of Plücker and Hittorf, and related to other observations by Goldstein, Spottiswood and Moulton, and others, about the same period.

In 1891 Sir William Crookes was President of the Institution of Electrical Engineers, and in his inaugural address he expounded further some of these brilliant experimental investigations, to which Schuster

and many others had contributed.   It is not too much
to say that up to the time of Crookes the phenomena
of the vacuum tube were shrouded in darkness, notwithstanding much laborious and painstaking work
done both in this country and on the Continent in
connection with them ; but that since the researches
of Crookes in the seventies, the theoretical luminosity
of the vacuum tube has steadily increased, until now,
as Maxwell predicted, it is shedding light upon the
whole domain of electrical science, and even upon the
constitution of matter itself.

CHAPTER III.

FORESHADOWING OF THE ATOM OR INDIVISIBLE UNIT OF ELECTRICITY.

So far we have dealt with the fundamental laws of electricity in general. It is now time to begin to consider the possibly atomic or molecular condition in which it is associated with atoms of matter.

Quoting again from the great Treatise of Clerk Maxwell, 1st Edition (1873), we find on page 312, in the chapter on electrolysis, the following sentence :

"Suppose, however, that we leap over this difficulty by simply asserting the fact of the constant value of the molecular charge, and that we call this constant molecular charge, for convenience in description, one molecule of electricity." . . .

Thus some idea of the conception of the atomic nature of electricity was long ago forced upon men of genius by the facts of electrolysis and a knowledge of Faraday's laws. But Maxwell went on, after a few more paragraphs :

"It is extremely improbable that when we come to understand the true nature of electrolysis we shall retain in any form the theory of molecular charges, for then we shall have obtained a secure basis on which to form a true theory of electric currents, and so become independent of these provisional theories."

It is rash to predict what may ultimately happen, but the present state of electrical science seems hostile to this latter prediction of Maxwell. The theory of molecular charges looms bigger to-day, and has taken on a definiteness, largely as the outcome of his own work, that would have pleased and surprised him.

The unit electric charge, the charge of a monad atom in electrolysis, whatever else it is, is a natural unit of electricity, of which we can have multiples, but of which, so far as we know at present, it is impossible to have fractions.

I will extract the following sentence from Section 32 of my little book called *Modern Views of Electricity* (1889. See also *Brit. Assoc.*, Aberdeen, 1885, p. 763) :

> " This quantity, the charge of one monad atom, constitutes the smallest known portion of electricity, and is a real natural unit. Obviously this is a most vital fact. This unit, below which nothing is known, has even been styled an ' atom of electricity,' and perhaps the phrase may have some meaning. . . . This natural unit of electricity is exceedingly small, being about the hundred-thousand-millionth part of the ordinary electrostatic unit; or less than the hundred-trillionth of a coulomb."

The atom with its charge is called an " ion." The charge considered alone, without attending to its

atom, was called by Dr. Johnstone Stoney an
" electron" or natural electrical unit.

What we learn with great accuracy from electrolysis is the ratio of the charge to the mass of
substance with which it is associated. It matters
nothing how much substance is chosen, whether 100
atoms or one, whether an atom or a gramme or a ton,
—the amount of electricity associated with it in
electrolysis, and liberated when the substance is
decomposed, increases in the same proportion; the
ratio is constant for each material, and if determined
for one is known for all.

This ratio is the reciprocal of what is technically
known as the "electrochemical equivalent" of a substance. In the light of Faraday's laws, if this quantity
is measured for one substance it is known for all,
because the charge is the same for every kind of
atom, up to a simple multiple; and hence in specify-
ing electrochemical equivalents there is nothing
to consider but the-atomic weight, or combining
proportion, of the substance. Thus the electrochemical equivalent of oxygen is 8 times that of
hydrogen, that of zinc is $32\frac{1}{2}$ times, and that of silver
108 times that of hydrogen. The substance chosen
for a determination of the electrochemical equivalent
may be the one which can be most accurately experi-
mented on; and Lord Rayleigh has shown that such a
substance is nitrate of silver, and has ascertained that
if a current of one ampere is passed from a silver anode
to a platinum cathode through a nitrate of silver solution, the cathode gains in weight by 4·025 grammes
every hour. Hence the electrochemical equivalent of
silver is

$$\frac{4\cdot025 \text{ grammes}}{1 \text{ ampere-hour}};$$

the electrochemical equivalent of hydrogen, being $\frac{1}{108}$th of this quantity, is

$$\frac{4 \cdot 025 \text{ grammes}}{108 \text{ ampere-hours}} = \frac{4 \cdot 025}{108 \times 360} \text{ c.g.s.}$$

$$= 0001035 \text{ c.g.s.} = \frac{1}{96600} \text{ grammes per coulomb.}$$

Hence the ratio of an atom of electricity to an atom of hydrogen is $9,660 \, \mu^{-\frac{1}{2}}$ c.g.s. units, or approximately

$$10^4 \sqrt{\left(\frac{\text{centimetres}}{\mu \text{ grammes}} \right)};$$

the unknown constant μ necessarily making its appearance, because we are comparing quantities of different nature, or at any rate quantities measured in different ways, viz., 'electricity' and 'matter' (see Appendix D).

The numerical part of this quantity is known with comparative exactitude,* that is to say up to the limits of error of experiment: to proceed further, we must make an estimate of the mass of an atom. That can be done, and has been done, in many ways, and we have been taught both by Dr. Johnstone Stoney and by Loschmidt, originally even by Dr. Thos. Young, but with greatest force and range by Lord Kelvin, that the mass of an atom of water is approximately 10^{-24} of a gramme; wherefore an atom of hydrogen will be approximately 10^{-25} gramme; whence the unit of electric charge is 10^{-21} c.g.s. magnetic unit, or 10^{-10} of an electrostatic unit or 10^{-20} of a coulomb.

* The decimal places are correctly printed above ; though the fact that 1 coulomb, or 1 ampere-second, is one-tenth of a c.g.s. unit—owing to the ohm and volt having been inadvertently defined, one as 10^9, and the other as 10^8 c.g.s., instead of both the same—always stands ready to introduce confusion and error.

I have emphasised this matter of the ratio m to e, or e to m, because it plays a considerable part in what follows. The absolute values are of less consequence to us than the ratio, and are only known approximately, but the ratio is known with fair accuracy; and the ratio $e : m$ for hydrogen is very nearly 10^4 magnetic units, or more exactly 9,660.

Thus what we learn from electrolytic conduction, briefly summarised, is that every atom carries a certain definite charge or electric unit, monads carrying one, diads two, triads three, but never a fraction; that in liquids these charges are definitely associated with the atoms, and can only be torn away from them at the electrodes; that the current consists of a procession of such charges travelling with the atoms,—the atoms carrying the charges, or the charges dragging the atoms, according to the point of view from which we please to regard the process.

CHAPTER IV.

FORESHADOWING OF THE ELECTRON.

*Separate Existence of the Electric Unit suggested
by Conduction in Gases.*

WE will now leave liquids and proceed to conduction
by rarified gases, that is to say to the phenomena
seen in vacuum tubes. If a long glass tube, say a
yard long and two inches wide, with an electrode at
each end, and full of common air, is connected to an
induction coil and attached to an air-pump,—the
ordinary spark-gap of the coil being, say, two or
three inches wide,—we find that for some time after
working the pump the electric discharge prefers the
inch or two of ordinary air to a long journey through
the partially rarified air in the tube ; but that at a
certain stage of exhaustion, one which any rough
air-pump ought to reach, this preference ceases. A
flickering light appears in the tube, readily visible in
the dark, which very soon takes on the appearance of
red streamers like the Aurora Borealis ; and soon the
sparks outside in the common air cease, showing that
the rarified air is now the better conductor and the
preferable alternative path. Let the exhaustion
proceed further—the axis of the tube becomes
illumined with a glow, which is now much brighter,

forming a band or thread of light, while the original
spark-gap may be shortened down gradually to one-
eighth of an inch, or even less, without any spark
taking place across it,—showing that rarified air is
a very good conductor. When the best conducting
stage is reached the tube is filled with a glow, called
the positive column ; and both ends of the tube are
apt to look alike. If we exhaust still further—and
to exhaust even as far as this something better than
an ordinary air-pump is necessary, an oil or mercury
pump being the most suitable—the column of light
is seen to fill the whole tube, to gradually lose its
bright red or crimson tint, and to break up into a
number of very narrow discs, like pennies seen edgeways. At the same time the spark-gap must be
widened to something more like a quarter or half an
inch, to prevent the discharge from taking that path,
and a dark space near the cathode now begins to be
visible, the cathode itself being covered all over with
a glow, while the anode is usually only illuminated
at a point or two. The striæ, into which the positive
column has been broken up, thicken and separate as
exhaustion proceeds. The dark space near the
cathode also enlarges, driving as it were the positive
column before it into the anode, and looking as if it
would presently fill the tube ; but before it can do
this it is noticed that the glow on the cathode itself
is coming off as a kind of shell, leaving another dark
space, a narrower and much darker space, inside it.
The first dark space has been called Faraday's dark
space ; the second is generally known by the name of
Crookes'. This second dark space now increases in
thickness, pushing the glow before it as the vacuum
gets better and better ; but the terminals of the
spark-gap must now be pulled still further apart, else

the discharge will prefer to take a reasonably long
path through the air.   Exhausting further still, the
glow all disappears and the second dark space fills
the whole of the tube; and now is noticed a new
phenomenon, the sides of the glass have begun to
glow with a phosphorescent light, the colour of the
light depending on the kind of glass used, but
generally in practice with a greenish light; a result
evidently of being the boundary of the dark space.
If exhaustion proceeds further, the resistance of the
tube becomes very high, and the spark may prefer to
burst through an equal, and ultimately even a greater,
length of ordinary air.   This is the condition of the
tube so much investigated by Crookes, by Lenard
and Röntgen, and by many other observers.   It is
the phenomena occurring in this dark space which
have proved of the most intense interest.

Cathode Rays.

So far we have supposed that the cathode is a
brass knob or other convenient terminal introduced
into the tube; but if we now proceed to use other
shapes,—as Plücker did first in 1859, followed by
Hittorf (1869), Goldstein (1876), and Crookes (1879)—
using a flat disc or a saucer-shaped piece of metal, and
if we then introduce into the dark space various substances, we shall find that shadows are thrown, and
that the dark space is full of properties which are
most clearly expressed by saying that it is a region
of cathode rays—that is to say, of something shot
off in straight lines from the cathode.   There is
evidently something being thus shot off—though whatever it is, it is invisible until it strikes an obstacle—
something which seems to fly in straight lines and to
produce a perceptible effect only when it stopped.

Such a 'something' might be a bullet from a gun, which is quite invisible when looked at sideways, but may produce a flash of flame when it strikes a target, or may do other damage. So it is with these cathode rays : the region of their flight is the dark space ; the boundaries of that space, where the projectiles strike, are illuminated. A substance with phosphorescent power, such as many minerals, or even glass, phosphoresces brightly ; and the path of the rays can be traced by smearing a sheet of mica with some phosphorescent powder and placing it edgeways along their path. In this way it can be shown that the rays are like particles travelling definitely in straight lines, not colliding against each other, but each shot like bullets from an immense number of parallel guns. Where they strike the sides of the glass they make it phosphoresce ; where they strike residual air in the tube, as they do if the exhaustion is not high enough, they make it phosphoresce also, and give, in fact, the ordinary glow surrounding the dark space.

These rays possess a considerable amount of energy, as can be shown by concentrating them, by means of a curved saucer-shaped cathode, and bringing them to a focus. The rays can be brought to a focus in consequence of the fact that they are projected from the cathode initially *normal* to its surface, though the focus is further from the cathode than the centre of curvature because of something equivalent to mutual repulsion of the rays. A piece of platinum put at that focus will (if the exhaustion is not too high) show evident signs of being red-hot—that is to say, will emit light. If the exhaustion proceeds further, less heat is produced, though a phosphorescent light is now emitted from suitable substances, like alumina and most earths; and if the exhaustion is pressed further still, the bombarded target emits no visible light but only that higher kind of radiation known as Röntgen or X-rays. It may be doubted, however, whether the target itself emits these rays, whether its function is not rather to stop the projectiles, as suddenly as possible, by the massiveness of its atoms. Thus the best target would be a substance with the heaviest atoms. X-rays are emitted by the suddenly stopped projectiles, in a manner which has been investigated both by Sir G. Stokes and Professor J. J. Thomson, and which is intelligible to anyone who has studied the properties of moving electric charges moving at or near the speed of light : a matter on which Mr. Heaviside has written with great clearness in his volume called *Electromagnetic Theory*.

Cathode rays have a remarkable penetrating power; for Hertz found that a thin metal diaphragm, especially if it were of aluminium, was powerless to stop their passage completely; as could be demonstrated by the phosphorescence and other effects appearing in the further half of the tube beyond the diaphragm.

The position of the anode in such experiments is of small consequence. There must be one somewhere, and the easiest plan is to make it a cylinder through which the cathode ray bombardment goes. The bombarding particles fly in straight lines and decline to turn a corner, taking no apparent notice of the position of the anode, and exhausting themselves by bombarding the side of the glass opposed to them; as can be well shown by having the tube bent into a V shape, for instance.

Lenard extended Hertz's discovery in a remarkable way by skilfully constructing a tube with an outer window of very thin aluminium, so arranged as to be able to stand the atmospheric pressure outside. He then directed the cathode ray bombardment on to this window or aluminium film, and showed that the rays can penetrate it and actually come outside into the ordinary atmosphere, where they are called Lenard rays, in honour of this indefatigable investigator, a friend and disciple of Hertz. (See Fig. 1.)

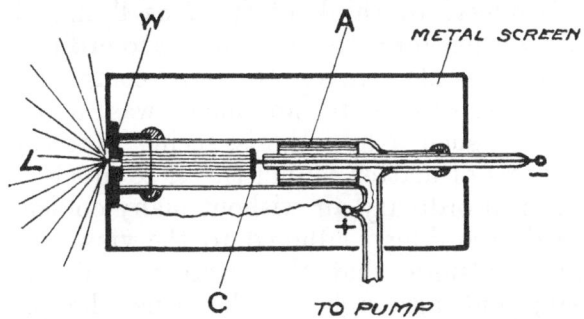

FIG. 1.—Lenard tube for the production of Lenard rays, which were discovered before Röntgen rays. C is a cathode in high vacuum; the anode A is a metal cylinder behind it; the whole is screened in metal, and the cathode rays impinge on a minute hole W covered with exceedingly thin aluminium foil, through which it would seem the rays emerge into the air, radiating in all directions from the aluminium window as Lenard rays L, where they are rapidly diffused and absorbed.

These Lenard rays make the air phosphoresce and produce the other effects which cathode rays can produce, but they are stopped within a moderate range by the immense obstruction they meet with from a substance of the density of ordinary air. Substances seem to stop them simply in proportion to the quantity of matter which they encounter, without regard to its nature. A thick layer of air would be about as opaque as a layer of water $\frac{1}{800}$ as thick; and even if the body put in their way is an opaque solid such as a sheet of metal, provided it is thin enough and not too massive, it will be penetrated by the rays; and phosphorescent effects will be produced on the other side of it. The rays can also affect photographic plates, and indeed do nearly all the things, though on a smaller scale and with much less penetrating power, that the later discovered Röntgen rays can do.

The Lenard rays are clearly cathode rays emerged from the tube; and it was the custom, at the date of their discovery, to think of them as flying charged particles of matter; though the extraordinary distance they could travel through common air, a distance comparable to an inch, was a manifest objection to such a hypothesis, seeing that things as big as atoms of matter cannot travel so much as $\frac{1}{1000}$ of an inch in ordinary air without many collisions.

Lenard accordingly adhered to the view, advanced first by Goldstein, that they were not material but ethereal; and although, in the sense he probably intended, this is not a tenable view—for they are not ethereal waves or anything of the nature of radiation—yet, as we shall see, neither are they ordinary material particles, any more than the cathode rays are. But that is just the point which we are now considering, and we will return to them as observed by Crookes in 1879.*

Nature of the Cathode Rays.

We have seen that the impact of the cathode rays, speaking in language appropriate to the assumption

---

* The biographical history of this subject is set out largely in the contemporary letters that passed between Crookes and Stokes: these have been supplied by Sir William Crookes, and will shortly be published in the *Scientific Correspondence of Sir George Stokes.*

that they are charged particles, will result partly in heat, or vibration of the impacted molecules ; partly in light or phosphorescence, due to the quiver of electrically charged atoms (or rather of electrical charges on atoms) as in the ordinary process of radiation ; and partly in X-rays :—all of which effects are readily seen at different stages of vacuum in a Crookes' tube. The *momentum* of the flying particles shot off from the cathode can also be exhibited by putting into their path some form of vane or little windmill, which will then be driven mechanically, as the vanes of a radiometer are driven by the recoil of the molecules of the residual air from the warmer surface,—a stress being thus set up between the vanes and their glass enclosure. In the electric vacuum tube experiment, the stress seems to be between the cathode, or gun, and a layer or stratum of the residual gas not very far from it—for unless the exhaustion is very high the gradient of potential close to the cathode is very steep,—so the propelling force is clearly the force of electrical repulsion, the particles travelling down the grade of potential just as they travel in ordinary electrolysis, and then proceeding for the rest of the journey by their acquired momentum. But whereas in ordinary electrolysis they meet with constant encounters and therefore progress very slowly, in a high vacuum they can fly for several inches in a free path without encountering anything, and therefore without causing any disturbance,—thus giving rise to no appearance but that of the dark space. Phenomena occur only where they strike.

This was the view of the nature of cathode rays taken by the whole world after Crookes' demonstration ; it was supposed that they were flying atoms, and that

they were flying with ordinary molecular speed, but
with a long free path—much longer than would have
been expected from ordinary gaseous theory. The
extraordinary length of free path was somewhat
difficult to reconcile with the doctrine that they were
flying atoms obedient to the ordinary laws of gases ;
except that, being subject to electrical propulsion all
in the same direction, their course was more regular,
and their encounters therefore fewer, or less effective
in causing deflection, than if they had been moving
at random. This same feature of regularity it is
which confers momentum upon them ; their motion
does not constitute heat, and is not to be considered
as corresponding to temperature ; they are moving
in orderly succession like an army or like a wind,
rather than with the irregular unorganised motion
appropriate, and solely appropriate, to the terms
" heat" and " temperature," and to the ordinary
kinetic theory of gases. Crookes indeed hazarded the
surmise—by one of those flashes of intuition which
are sometimes vouchsafed to a discoverer but are often
ridiculed by representatives of orthodox science at the
time—that he had obtained matter in "a fourth
state ;" and even that he had got in his tube something equivalent to what was contemplated in the
" corpuscular" theory of light. There is some correspondence with fact even in this last mode of
statement, when the particles are moving quickly
enough, for a nuclear wave-front or ether-pulse is
then travelling with them ; but how true the first
was—that the matter in the dark space was in a
fourth state, neither solid nor liquid nor gaseous—
how true that was we shall presently see.

Meanwhile let us summarise the evidence for the
view that the cathode rays are at any rate charged

particles of some kind, in extremely rapid motion.
That they are in motion must be granted from the
fact of their bombardment—driving mills, heating
platinum, and the like ; and in order to show that
they are charged, the most direct plan is to catch
them in a hollow vessel connected with an electroscope, as Perrin did ; but another plan is to show
that they have the properties of an electric current.
If they are charged while in motion they constitute a
current, on Maxwell's theory, and therefore should
be able either to deflect a magnet or to be deflected
by it; and here comes one of the most simple and

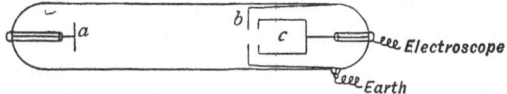

FIG. 2.—Simplest form of Perrin's apparatus for proving that cathode
rays carry a negative charge. The rays from *a* pass through an earthed
screen *b* into a hollow or Faraday vessel *c*.

important experiments in physics at the present
time. A definite form of some old experiments
foreshadowed by Plücker (1862), and developed by
Hittorf, Goldstein, and many other vacuum tube
observers, was arranged by Crookes in 1879, when
he made the track of the rays visibly luminous by
passing a pencil of them through a slit and letting
them graze along the surface of a film of mica covered
with phosphorescent powder, and when he then
brought near them a common horseshoe magnet.
When this is done the track of the rays is at once
seen to be curved ; showing that it is not a beam
of light we are looking at, but a torrent of charged
particles ; since they behave like an electric current
and are deflected by a magnet. It is ultimately the
very same phenomenon as can be observed with
difficulty, owing to its smallness, when a current

L. E.   C

flows through metals,—an effect discovered by E. H. Hall in America, and known as the Hall effect.

The fact that the particles are thrown off the cathode, being evidently vigorously repelled by it, is sufficient to suggest that they must be negatively charged; the direction of the curvature caused by a magnetic field enables us to verify at once that the flying particles are *negatively* charged, and no comparable rush of positive particles in the opposite direction, or in any direction, has been observed. The speed of transmission of the positive *current* is very great, and it must be conveyed by a multitude of positive particles, but individually their motion is comparatively slow (see however chap. vi.). In that respect evidently the magnetic curvature of cathode rays in gases differs from the magnetic curvature of a current in metals; viz., that whereas in metals the major action is sometimes upon the negative and sometimes upon the positive stream, according to the nature of the metal—the difference, which is all that is observable, being always small,— in the cathode stream it is the negative alone that is acted upon, and so the action is always large.

It seems, therefore, that for some reason or other the negatively charged bodies in a vacuum tube are much more mobile than the positive, and that the mobility of the negatively charged bodies is extreme. One striking method by which their mobility was displayed consisted in the fundamental observation by Professor Schuster * that all parts of gas in a closed vessel became conducting when an electric discharge had taken place in one corner of it; so that even though the vessel consisted of different compartments, one compartment was made feebly

* Bakerian lecture 1890, *Proc. Roy. Soc.*, vol. 47, p. 526.

conducting by a discharge in the other, provided
that the two had any kind of gaseous communication; a fact which looked as if some extremely
mobile particles, probably the negatively charged
particles of cathode rays, could wander about to
a considerable distance in a very short time and
take their share in the conveyance of an electric
current. The conductivity of gases appeared to be,
indeed, entirely due to these loose or dissociated or
detached charged particles, or ions, and where they
were absent the gas did not conduct at all; it
could be broken down, being a weak dielectric, by a
sufficiently strong force, but it would not leak;
whereas, when these loose charged particles were
about, it leaked readily, becoming to all intents and
purposes an electrolyte amenable to the feeblest
electric influence. The production of this electrolytic
condition is called " ionisation." The act of breaking
down the air by an electric discharge was thus
found to render the surrounding air for a time
electrolytic. Its electrolytic quality, however, did
not last long. The mobility of the particles which
enabled them to travel to a considerable distance
also enabled them to get rid of themselves by
clinging to the sides of the vessel, or perhaps by
re-uniting with some opposite charges, which after
some time in their rapid journeys they must
casually encounter. Prof. Townsend,* however, found
that the conducting power lasted unexpectedly long
if no dust was present : the dust particles apparently
acting as intermediate receivers and storers of
charge, promoting interchanges, which did not very

* J. S. Townsend of Trinity College, Dublin, then working in the
Cavendish Laboratory, Cambridge, now Waynflete Professor of Physics
in the University of Oxford.

readily occur through direct encounters. And the time that thus elapsed before the whole of the conductivity disappeared from dust-free air suggested that the moving particles must be very small, so that intimate collisions were comparatively infrequent.

The mobility or diffusiveness of the molecules of a gas depends on their mean free path, and that depends on their atomic size; the smaller they are, the more readily can they escape collision. Hence it is that collisions are so rare in astronomy : the bodies are small compared with the spaces between them. The behaviour of charged particles seemed to indicate that they must in some cases be something smaller than atoms : it seemed hardly likely that material atoms could behave in the way they did. It was recollected that it had occurred to some philosophers, among them Dr. Johnstone Stoney, that electric charges really existed on an atom in concentrated form, not diffused all over its surface but concentrated at one or more points,—perhaps acting as satellites to the main bulk of the atom; so on that view it was just possible that these flying particles might be not charged atoms at all, but charges without the atoms, the concentrated charges detached—knocked off as it were in the violence of the discharge, and afterwards going about free. Such particles would naturally travel at an immense pace, because they would still be exposed to the full electric force that they had experienced before, and yet would have shaken off the encumbrance of the material atom with which they had been associated. It is true that no such disembodied charges, or electric ghosts, had ever been observed. All the experiments that had been made in electrostatics had been made on charged matter,—the surface or boundary of the matter acting as the locality for an electric charge ; and no other locality for a charge was known. The facts of electrolysis had suggested or proved that the atoms themselves could carry charges, and hence that if a liquid were electrified, it must be due to some of the atomic charges of one sign, appearing in overbalancing proportion at the surface ; though perhaps still in association with their respective atoms.

Yet at the same time the occurrences at an electrode, where an ion plainly gave up its charge and escaped without it, indicated the *possibility* that perhaps the electric charge could exist alone ; at any rate that it could be handed from one atom to another, and thus might conceivably exist alone for an instant. During this momentary isolation some charges might, in the freedom of a rarefied gas discharge, possibly escape, and wander about free.

To such hypothetical isolated charges, the unit charge or charge of a monad atom, the name " *electron* " had been given ; and when I speak of an " electron " I mean to signify the, at first purely hypothetical, isolated electric charge. Whereas by the term " ion " I always signify the atom and its charge together. The ions consist of Faraday's anions and cations. Lord Kelvin prefers the term electrion to electron.

Now if the flying particles which constitute the cathode rays were electrons rather than ions,—if they were detached charges, leaving their atoms behind them (necessarily leaving those atoms positively charged),—their extreme mobility and diffusiveness and high speed would be perfectly natural ; and although they would not be 'matter' in the ordinary sense, yet no difficulty need be felt at their possessing

some of the properties of matter, at any rate such
properties as appertain to matter by reason of its
having inertia ; because, as we have seen, an electric
charge itself does possess a certain kind of imitation
inertia.   Hence these electrons in movement would
possess momentum, and might therefore propel windmills (though the actual motion of the windmills in
Crookes tubes seems more likely due to charge and
electric repulsion than to simple momentum) ; they
would possess kinetic energy, and therefore might
heat a piece of platinum ; and if suddenly stopped by
a massive target when travelling at a high speed they
might readily give rise to phosphorescent appearances,
and even to the sudden pulse of radiation known as
X-rays.   Indeed the existence of this last property is
capable of clear deduction on electrical principles
if the matter is further gone into.   (See chap. ix.)

   The continued passage of a current through a
vacuum tube cannot be explained by a torrent of
electrons alone,—there must be some mechanism
for continually producing them fresh and fresh,
near or at the cathode, else they would almost instantaneously get exhausted.   The most probable
view of the matter is that suggested by J. J. Thomson :
that the current is conveyed chiefly by positive ions,
which are produced in the residual gas by ionisation
due to the first discharge of cathode rays.   These
positive ions then pass along comparatively slowly
toward the cathode, creep in towards it, as best they
can, in face of the bombardment ; and then at the
last—experiencing the violent gradient of potential
in its immediate neighbourhood—rush up against it
and by their shock produce a fresh supply of electrons.
The glow over the cathode is supposed to mark the
region of this ionisation.   The negative particles, thus

set free, then fly off as cathode rays, setting up fresh
ionisation, and producing a copious further supply of
positive ions ;—on the existence of which the possibility of the cathode rays themselves depend. The
positive and the negative particles on this view are
mutually dependent : each is the cause of each ; and
when either fails to be formed in a vacuum tube it is
impossible for it to conduct, even when its terminals
are highly electrified; for if the supply of either sign
of ion is stopped, that of the other at once fails.
This accounts for the action of ' electric valves,' wherein
the positive ions are prevented from getting at the
cathode in one direction, by reason of a special
arrangement for concentrating an electron bombardment along the direct route, without any back door
or side entrance for the positive ions. The provision
of such a back door, even though the route thereto
be long, immensely eases the conveyance of current :
as was strikingly shown by Hittorf.

It has been observed that any obstacle introduced
into the dark space near the cathode, if it is able to
check locally the access of positive ions, will throw a
shadow both fore and aft,—one towards the cathode,
and likewise one down the cathode rays,—because
the generation of fresh electrons is thereby locally
prevented.* At this stage we may conveniently
summarise the position thus :—

The magnitudes which need experimental determination in connexion with cathode rays, in order
to settle the question and determine their real
nature, are the speed, the electric charge, and if
possible the mass, of the flying particles.

Everything suggests that they are flying with

* Schuster, *Proc. Roy. Soc.*, xlvii., p. 557, 1890 ; Wehnelt, *Wied. Ann.*
lxvii., p. 421, 1899.

prodigious speed, but it is desirable to make a measurement of that speed.

The force of propulsion exerted on them indicates that they are highly charged; and their penetrating power suggests that they are excessively small, so that to them ordinary solids, such as metal sheets, appear porous; but an experimental method is necessary to determine what may be called their electrochemical equivalent,—that is to say the ratio of their mass or inertia to their electric charge,—even if it be not possible to determine the mass and the charge separately.

In electrolysis the electrochemical equivalent, or the ratio m/e, depends on the nature of the substance; and for hydrogen is of the order 10^{-4} in electromagnetic units, as stated in Chapter III. It is a matter of great importance to determine the value of the same ratio for the cathode rays, and to ascertain whether it varies with the substance contained in the vacuum tube, or whether it is the same for all substances—being characteristic of a single variety of the flying particle and of nothing else.

CHAPTER V.

DETERMINATION OF SPEED AND ELECTROCHEMICAL EQUIVALENT OF CATHODE RAYS.

If the cathode rays consist of flying electrified particles they will be deflected, or their paths curved, by the proximity of a magnet: and this is a well known and prominent fact concerning them. With some care the amount of deflexion, caused by a magnetic field of known strength, may be measured.

The curvature of path produced in cathode rays by a transverse magnetic field, or the amount of spirality produced by a longitudinal magnetic field, constitutes an evident mode of attacking the problem of estimating their velocity.

If the velocity is constant and the magnetic field uniform, the curve into which the stream is bent round the lines of force will manifestly be a circle; and its course can be readily traced either directly, after Crookes' manner, by letting it graze a phosphorescent substance, or indirectly by inference from the position of a linear target placed so as to catch the deflected rays. If the direction of velocity is inclined to the direction of the field, the course of a particle will be compounded of a circular motion round a line of force, and an unchanged rectilinear motion along it: that is to say, it will be a spiral, more or less elongated, threading itself along the negative field : the direction of twist depending on the sense of the field.

There is no difficulty in determining the radius of curvature r ; and the theory of normal deflexion is the simplest possible,—nothing more than stating that the magnetic force H, acting on the current element eu, is the necessary deflecting or centripetal force, mu^2/r, required to overcome the mechanical inertia of the particles ; $i.e.$,

$$\frac{mu^2}{r} = \mu eu \mathrm{H},$$

whence $\dfrac{m}{e}u = \mu \mathrm{H} r$;

or the ratio e/m is to the velocity of the particles as the curvature of their path is to the intensity of magnetic field which curves it.   Prof. Schuster of Manchester was among the first to make measurements of this kind.

The two factors on the right of this equation are directly measurable (μ being conventionally ignored as usual, or—a better mode of expression—united with H as induction-density) ; but the two factors on the left are both unknown, hence neither can be determined by this means alone,—an assumption must be made about one or other of them, or else another independent kind of experiment must be made.

Assume, as many experimenters did, that u is a velocity appropriate to atoms flying in a gas of ordinary temperature, then the value of e/m comes out not so very far discrepant from the usual ionic value, measured in liquid electrolysis, viz., 10^4 c.g.s. Or, conversely, assume the usual ionic or electrolytic

value for this ratio, and the cathode ray velocity
comes out something quite appropriate to atoms
of matter.

This, however, is a trap. These accidental coin-
cidences may retard progress in a most serious
manner, for they satisfy the mind and deter people
from investigation. It is almost impossible to be

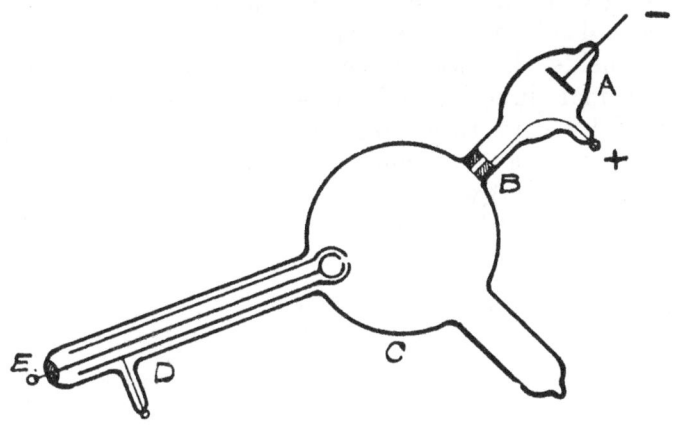

FIG. 3.—Modified Perrin apparatus adopted by J. J. Thomson for
measuring the charge, and at the same time the magnetic deflexion, and
sometimes the thermal energy also, of cathode rays. The rays from the
cathode, after passing through a perforated anode and proceeding in a
straight line, can be deflected by a magnet a measured amount, so as
just to enter a hole in an earthed guard-screen D, and then a hollow
cavity provided with an electrode E, whereby the aggregate charge
conveyed by them is measured.

completely on guard against them, and they are
usually accepted until a more thorough qualitative
acquaintance with the subject leads to an instinctive
feeling that something is wrong somewhere.

So it was in this case : the long free path and the
penetrating power of the cathode rays kept insisting
that the particles were not really atoms of ordinary
matter,—a truth which both Lenard and Crookes
had instinctively grasped, in spite of much criticism

and valid arguments the other way; so in 1897 J. J. Thomson made a much more serious attack on the whole position.

He arranged that the magnet should deflect the rays into an insulated hollow vessel, connected with an electrometer and a known capacity, so that the aggregate charge of the cathode ray particles collected in a given time could be measured by the rise of potential observed (cf. Fig. 3). He also arranged that inside the hollow vessel they should fall upon a thermal junction of known heat capacity, connected by very thin wires to a galvanometer (acting therefore as a calorimeter), so as to measure their aggregate energy.

Thus he could make the following simultaneous determinations :

$$Ne = Q$$
$$N\tfrac{1}{2}mu^2 = W$$
$$\frac{m}{e}u = \mu Hr.$$

In these three equations there are four unknown quantities; but one pair can be treated as a ratio, and another, N, can be eliminated, and thus we get—

$$u = \frac{2W}{QHr}.$$
$$m/e = \frac{Q}{2W}(\mu Hr)^2.$$

When these brilliant measurements were actually made in the laboratory, the atomic nature of cathode rays was, if not actually disproved, at all events rendered highly improbable; for their speed was found to be of the order ten thousand miles per second, or even as high as $\frac{1}{10}$ that of light in a favourable case, being always of the order 10^9 c.g.s., while the electrochemical equivalent was of the order 10^{-7} c.g.s., or about $\frac{1}{1000}$ that of hydrogen.

Changing the kind of residual gas in the tube, and changing the electrodes, made no difference to this last value. *The cathode rays were evidently independent of the nature of the matter present* : an exceedingly momentous fact. If they were matter at all, they appeared to be matter of some fundamental kind, independent of the distinctions of ordinary chemistry. Their velocity, however, depended on the potential difference between the electrodes, in a way that suggested that they were really projectiles urged by the potential gradient acting along a given length of path. They were propelled by the cathode through an aperture in the anode, and the measurement of their speed was made in the tube beyond the anode, where they are travelling by their own momentum. The distance apart of anode and cathode did not, and on the projectile hypothesis ought not to, affect this speed; for though the potential gradient is steeper when anode and cathode are put close together, the length of path during which the particles are subject to it is diminished by a compensating amount,—so that the velocity is theoretically independent of the distance between the electrodes, as long as the total difference of potential is maintained; it is the absolute difference of potential that determines the speed. (This is a familiar result of the conservation of energy, and is illustrated by ordinary falling bodies.) But manifestly if the electrodes are too close together it may be difficult to secure a high difference of potential between anode and cathode, since they may spark into each other outside the tube; and if there is much residual gas in the tube it will likewise be difficult to maintain a high potential difference; because that residual gas, under the influence of the cathode rays, will conduct. Consequently the best speeds are obtained at high vacuum; and if the density of the residual gas inside the tube is constant, the speed will be constant. The nature of the electrodes makes no difference, unless they give off gas or otherwise make it difficult to maintain the required potential difference.

Although the speed of the particles in cathode rays was thus found excessively great, their energy was only moderate, and their aggregate mass was therefore proved to be excessively minute; their aggregate electric charge, however, was considerable. They were able to raise an electrical capacity of 15 microfarad several volts, sometimes as much as 5 volts, in the course of a second; and in the same time they might be able to raise a calorimeter, whose heat capacity was about 4 milligrammes of water, by $2°$ C. Nevertheless their mass was so small that it would have taken one hundred years to collect a weighable amount: and then only about one-thirtieth part of a milligramme. They travelled with a velocity a hundred thousand times greater than the speed of rifle bullets, and represented the greatest velocity up to that time observed in matter,—if matter they were. And the electrochemical equivalent, instead of coming out in accordance with that observed in liquids, came out some thousand times smaller; that is to say, the charge associated with each particle of the cathode rays seemed a thousand times greater, in proportion to the mass, than the charge associated with an electrolytic ion, even of hydrogen.

If the flying particles were really atoms, there was no escape from the certainty that they were extraordinarily highly charged atoms; but if, as seemed more likely to the instinct of most of those who worked at the subject, the charge on the flying particles was the same as the charge possessed by an atom in electrolysis, then, assuming that the experiments were correct and correctly interpreted, there would be no escape from the conclusion that the mass associated with the ionic charge in cathode rays must be a thousand times smaller than the mass of a hydrogen atom; in which case the cathode projectiles might conceivably be the detached and hitherto hypothetical individual electrons or atoms of electricity themselves. It would be extremely rash, however, to jump to such a far-reaching conclusion on such comparatively scant evidence. The evidence must be confirmed by other departments of Physics or by other determinations based on a different method; and they must be further scrutinised in the light of other and totally different phenomena. We will first describe a determination made by another method, and then some striking confirmatory measurements applied to phenomena which belong apparently to other departments of Physics.

Further Measurements of Cathode Ray Velocity and m/e Ratio by Aid of Electrostatic Deflection.

Another and perhaps simpler method of determining the two quantities u and m/e was also employed by J. J. Thomson,—was indeed the first used by him, though it was not safe to draw a full deduction from it alone—viz., by deflecting the same rays both electrostatically and magnetically;

by introducing a pair of supplementary electrodes, one above and one below the course of the rays inside a vacuum tube, and connecting them to the poles of a low-potential battery,—a few storage cells for instance,—thus obtaining a vertical electrostatic field at right angles to the cathode rays. At the same time a magnetic field, produced by lateral magnet poles or by the lines of force due to an electric current in a circular ring, could be arranged at right angles to both the other directions; and thus the electrostatic deflection could be compared with, or could be used to neutralise, the magnetic deflection. The electric field, being fixed in direction and urging the particles along—not across—the lines of force, will act differently to a magnetic field and will cause the particles to move in a parabola—the shape which the earth gives to a horizontal jet of water.

Fig. 4 shows J. J. Thomson's apparatus for measuring the deflexion of cathode rays caused by an electric field at right angles to them.* The rays starting from the cathode C traverse a couple of earth-connected slits AB and after having passed between the electrified plates D and E, impinge on the glass at P, producing a small but vivid phosphorescent spot. The position of this spot is read on a scale pasted on the outside of the glass. In ordinary vacua screening prevents any effect from being observed, by reason of conducting power in the residual air, developed by ionisation caused by the impact of the flying particles; consequently Hertz failed to obtain the deflexion which was otherwise to be expected; but by pushing the vacuum to a higher stage J. J. Thomson overcame this difficulty, measured the deflexion PP', and employed this

* *Phil. Mag.*, vol. 44, p. 293 (1897).

method to assist in measuring the velocity and other constants of the cathode ray particles.

By noting the shift of the luminous spot, the deflexion of the narrow beam which has travelled through a length l of either an electric field of strength E, or a magnetic field of strength H, can be directly measured.

If u is the original velocity of the ray particles, travelling at right angles to one of the deflecting fields, either of them will have a time l/u in which

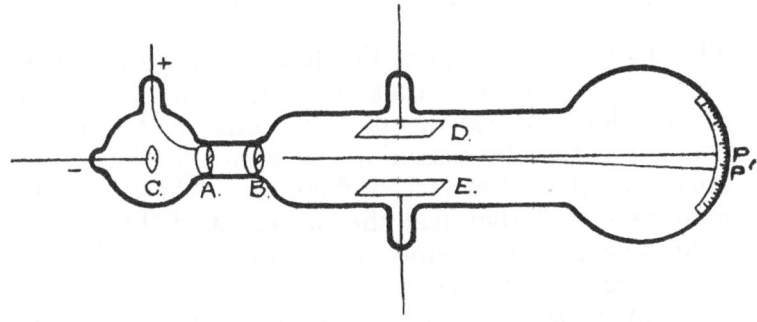

Fig. 4.—Thomson's apparatus for observing and measuring the electrostatic deflexion of cathode rays.

to act; and in that time an extra velocity w will be caused in the direction of the electric force, or perpendicular to the direction of the magnetic force, such that the rate of change of momentum of each particle will be $\dfrac{mw}{l/u} = \mathrm{E}e$ in the one case, and $= \mu \mathrm{H}eu$ in the other; wherefore the deflexion, if small, will be—

$$\theta = \frac{w}{u} = \frac{e}{m} \cdot \frac{\mathrm{E}l}{u^2} \text{ in the one case,}$$

$$\text{and } \theta' = \frac{e}{m} \frac{\mu \mathrm{H}l}{u} \text{ in the other.}$$

Hence
$$u = \frac{E}{\mu H} \cdot \frac{\theta'}{\theta}$$

and
$$\frac{m}{e} = \frac{l\mu^2 H^2}{E} \cdot \frac{\theta}{\theta'^2}.$$

This method, when applicable, appears to give fairly accurate results; and the outcome of the measurements is, that when H or CO_2 or Air is in the tube,

$$u = 2 \text{ or } 3 \times 10^9 \text{ centimetres per second,}$$

and $\frac{m}{e}$ = from 1·1 to 1·5 × 10⁻⁷ c.g.s. units.

The chief difficulty about this mode of experimenting is caused by the fact that the ionisation of residual air in the tube causes it to become a temporary conductor, thereby screening the flying particles from most of the electrical influence. There is no guarantee that they feel the full effect of the electric field which is ostensibly being applied; indeed it is not easy to let them feel any of the effect. It used to be thought that they were not susceptible to electrostatic action at all, and this was often adduced as an obvious argument against their being electrically charged particles; but fortunately Thomson soon surmised the cause of this masking of the simple effect to be expected, and succeeded in showing that with high enough vacua, and other precautions, the screening ionised atmosphere could be removed, and the electrostatic deflexion metrically observed.

Measuring Velocity by combined Electric and Magnetic Deflexion Method.

Now that it is possible to apply the electrostatic deflexion method to curve the path of flying charged particles, the simplicity of combining this with magnetic deflexion, and thereby making a couple of measurements simultaneously, is so great that it has practically replaced the more elaborate plan first described,—namely the method by observing the aggregate charge, the aggregate energy, and the magnetic deflexion only,—a method which, first as an original determination, and now as a check, has been of high value.

The simple plan has now been applied to rays of various kinds, and it may be well to give its simplest possible form, before proceeding to a more complicated case. It consists, (1) in observing the radius of curvature r caused by a magnetic field of measured strength H; (2) in finding the electric field E which, applied at right angles to the magnetic field, just succeeds in neutralising all deflexion; then a centrifugal force equation

$$\frac{mu^2}{r} = \mu e \mathrm{H} u,$$

$$\text{or} \quad \frac{m}{e} = \frac{\mu r \mathrm{H}}{u}$$

applies to experiment No. 1; and a balanced force equation

$$\mathrm{E}e = \mu e \mathrm{H} u,$$

$$\text{or} \quad u = \frac{\mathrm{E}}{\mu \mathrm{H}}$$

applies to experiment No. 2.

This second experiment gives the velocity, by itself; and the first in combination with it gives the electro-chemical equivalent.

So the determination of velocity for the case of particles flying all with one speed is remarkably easy: it comes out, in centimetres per second, as simply the ratio of the strengths of the electric and magnetic fields (both expressed in EM units, that is with $\mu = 1$) which can produce equal effects upon the flying particles, and which therefore, if applied in opposition, are just able to neutralise each other.

Note on Dimensions.

To verify that the "dimensions" of the last given equation are correct, we can remember that an electric field is of dimension $\dfrac{F}{e} = \dfrac{F}{l\sqrt{(\kappa F)}}$, where F is force, and l is length; and that a magnetic field is of dimension $\dfrac{F}{m} = \dfrac{F}{l\sqrt{(\mu F)}}$; so the ratio of an electric to a magnetic field is $\sqrt{(\mu/\kappa)}$; wherefore $\dfrac{1}{\mu}$th of that ratio is $\dfrac{1}{\sqrt{(\mu\kappa)}} = $ a velocity.

For practical purposes it may be convenient to write the equation as a proportion sum thus :

$$\frac{\text{the velocity of the particle}}{\text{the velocity of light}}$$

$$= \frac{\text{the electric field in electrostatic units}}{\text{the magnetic field in electromagnetic units}} ;$$

it being understood that the fields are adjusted till their effects are equal.

Effect on Lenard Rays.

Another mode of demonstration was employed by Professor Lenard, and served to identify his rays with escaped cathode rays.

Fig. 5 shows Lenard's apparatus for showing that his rays are accelerated or retarded by passing along the lines in an electric field between two disks *ab*. Here C is the cathode of a Lenard tube similar to that shown in fig. 1, and the rays enter an aluminium window at A, thence passing on through an earthed tube *n* leading to one of the charged disks. The disks are perforated to allow the rays to pass through them ; they go on through another tube

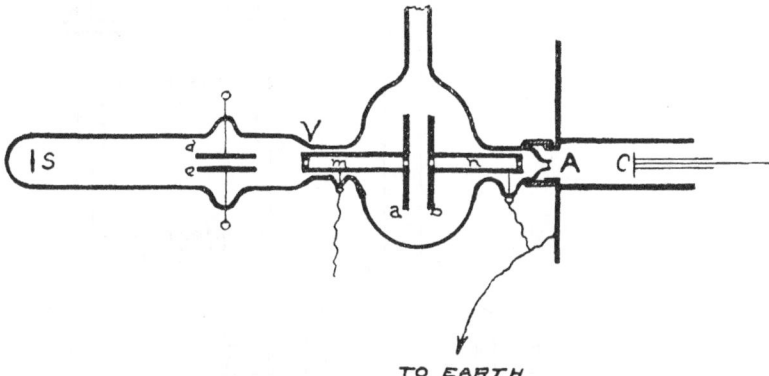

FIG. 5.—Lenard's apparatus for exhibiting the effect of a longitudinal electric field in affecting the velocity of cathode or Lenard rays.

m, which, with the disks, is kept at high potential, and thence between the plates *de*, which represent either an electric or a magnetic field ; after which they encounter a phosphorescent screen S, and depict upon it a luminous spot. The observation consists in observing the change of position of this spot when the sign of the electrification of the plates *ab* is changed ; for if the longitudinal field alters the velocity, the deflexion caused by the field *de* will also be modified, and thereby the change in velocity is demonstrated.

Direct Determination of the Speed of Cathode Rays.

Figs. 6 and 7 illustrate sufficiently the ingenious device applied by Wiechert, in accordance with a method of Des Coudres, for directly measuring the velocity of cathode rays during the time of their

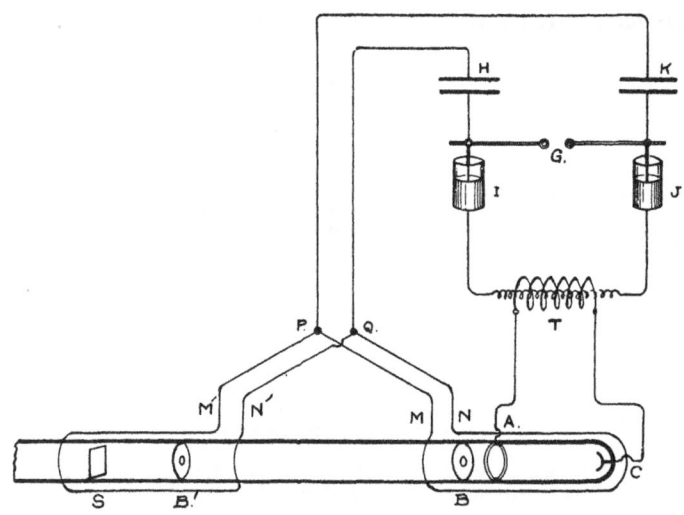

FIG. 6.—Connexions of the circuits in Wiechert's experiment for measuring directly the speed of cathode rays as they travel down a tube.

transit down the axis of an exhausted glass tube of moderate length.

Referring to either of the figures, the tube has a curved cathode C at one end, whereby the rays are focussed through an aperture in a screen B; they then travel down the tube to another perforated screen B', whence the central portion of them reaches a phosphorescent screen at S.

The tube is excited by a kind of Tesla induction

coil T, whose terminals lead respectively to the cathode C and to a ring-anode A.

Leyden jars I and J can be excited by a Ruhmkorff coil, or in any ordinary manner, and they discharge across the spark-gap G. Thereby a subsidiary oscillating circuit, of very high frequency, characterised by the condensers H and K, and limited to the region above the spark-gap G, is set in action; and an oscillatory discharge is conveyed to the symmetrical terminals PQ, whence it can pass round a pair of wire loops in parallel, MN

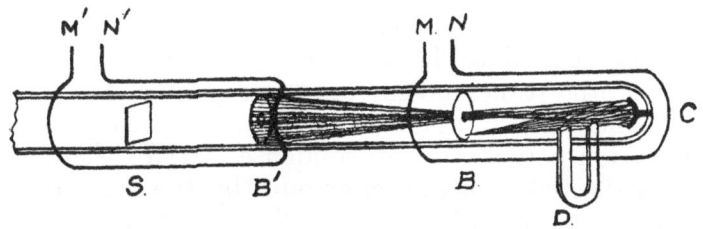

FIG. 7.—Enlarged portion of fig. 6, showing the vibration of the cathode rays by an alternating circuit and the adjustment of their range of excursion by a fixed magnet.

and M′N′. The effect of the alternating field, thus generated in the loop MN (fig. 7) is to wave the cathode rays rapidly to and fro, so that only in the middle of their path will they pass through the aperture in the disk B—the effect on the distant phosphorescent screen being then quite faint. But if a permanent magnet D is applied to them, the oscillation is deflected, and can be limited to the region between the centre and the circumference of the disk—as illustrated by the figure; thereby a greater number of rays get through, and the screen becomes more luminous than when the magnet is not applied, because the aperture will now be at the extremity of the swing of the oscillating rays. If

the magnet is too strong the screen S will become
dark, because the oscillating rays will be too much
deflected, so it is possible to tell pretty sharply the
phase of the oscillating rays in the right-hand portion
of the tube by watching the screen S and adjusting
the strength of the magnet.

So much for the right-hand end of the tube. Now
proceed to· the left-hand end, or rather to the
movable portions depicted to the left of figs. 6 and 7.
We have there also an oscillating current, $M'N'$, in
precisely the same phase of standing oscillation as
its other branch MN ; consequently, if the rays take
no time to travel down the tube, those which get
through the disk B' will be deflected as before ;
but if the time taken to travel down the tube
corresponds with a quarter of the extremely rapid
oscillation-period of the condensers HK, then the
rays deflected a maximum amount by the first branch
of the circuit will not be deflected at all by the
second, and therefore will reach the middle of the
screen.

By either altering the frequency of oscillation, or
adjusting the distance apart of the parts represented
by BB , it is possible to cause the deflexion on the
left-hand side to be either in the same phase, or in
opposite phase, with the first half ; or to be a zero, of
the first, second, or third order.

In this way, though obviously the experiment is a
difficult one, Wiechert was able to make measurements of the speed of the cathode rays produced
under given circumstances.

Unfortunately this speed is not a fixed and definite
constant, like the velocity of light: it does not indeed
depend upon the nature of the gas in the tube, nor
on the substance of the electrodes, but it does vary

with the density of the residual gas and with the intensity of the electric field.

In order to prevent the diffusion and spreading out of the rays, on their passage down the tube, a longitudinal magnetising helix was applied to it, so as to concentrate the rays along the axis.

If u is the velocity to be measured, and l the distance they travel to give the first zero, then $u = 4nl$, where n is the frequency of the electric oscillation in the circuit MN.

In one experiment l was 39 centimetres, and n was 32 million per second; wherefore u comes out $4\cdot5 \times 10^9$ cm. per sec.

As soon as the velocity is known, the ratio e/m can be determined by measuring the magnetic deflexion of the rays. The range of uncertainty for this determination, as made by Wiechert, lies between $1\cdot55$ and $1\cdot01$ times 10^7. It can hardly be considered the most accurate method of measurement, but its directness and ingenuity entitle it to attention.

CHAPTER VI.

DETERMINATION OF ELECTROCHEMICAL EQUIVA-LENT IN THE CASE OF ELECTRIC LEAKAGE IN ULTRA-VIOLET LIGHT.

THE same ratio of m to e, or a ratio of quite comparable magnitude, is obtained from phenomena which at first sight appear to be distinct.

One of these phenomena is the effect of ultra-violet light in discharging negative electricity from a clean metal or other surface ; a phenomenon discovered by Hertz, and the investigation of which was continued especially by Righi and by Elster and Geitel. (See one of the appendices to my " Signalling without Wires," published by the Electrician Co.) If ultra-violet light, whether from a spark or from a flame, fall upon a negatively electrified surface, then in general there will be a leak of electricity from that surface : which electricity can be received by any body placed opposite the illuminated one, and can be used to charge an electrometer of known capacity, and so be measured. The writer, assisted by Mr. Benjamin Davies, has made very many experiments in this subject, which, however, have not yet been published. Now Elster and Geitel made the notable discovery that the application of a magnet affected the rate of leak,

according to the direction of its lines of force. This
phenomenon suggested a magnetic deflexion of the
lines of leak, which were shown by Righi to be
singularly definite trajectories, and indicated that
the leakage was due to the bodily propulsion of
negatively electrified particles analogous to the
cathode rays. A vacuum is not necessary to observe
the effect, but in a vacuum the effect is more prominent and more accurately measurable. The
difference between this case and an ordinary vacuum-
tube case, is that there is no great E.M.F. or gradient
of potential applied, so there is nothing of the
nature of a disruptive discharge ; in fact there is no
leak at all until—by the stimulus of the presumably
synchronous vibrations of ultra-violet light—the
molecules are thrown into a state of agitation, and
the attachment of the negative charge, or of some
negatively charged corpuscles, thereby loosened.

Two things are necessary to get the particles away
from the plate ; they must be loosened by the impact
of ultra-violet light—the direction of polarisation of
this light having a very decided influence when
the surface is smooth,—and the surface on which
they exist must likewise be negatively charged,
so as to repel them. Neither light alone nor
electrification alone will produce any considerable
effect ; co-operation is necessary. Light alone is
able to cause a *slight* positive electrification,* by
the diffusion away of negative corpuscles—a process
which is assisted by a blast of air. Whether
electrification alone can produce a perceptible effect
depends on the temperature of the metal : when
that is high, it can—as was discovered by Guthrie.

* This effect was discovered simultaneously by Righi and by
Hallwachs. See *Phil. Mag.*, 1888, April, p. 314, and July, p. 78.

J. J. Thomson devised a most ingenious method of carrying out this experiment—that of discharge by means of ultra-violet light—in a metrical manner, and of deducing from it the electrochemical equivalent of the charged particles, that is to say the amount of matter which each contains compared with the electric charge which each carries.

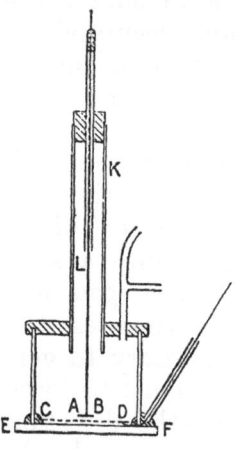

Fig. 8.—J. J. Thomson's apparatus for measuring the magnetic deflexion of the electric charge thrown off clean negative metal by ultra-violet light in a vacuum. The negative electrode is a clean zinc plate AB which can be raised or lowered, the other is wire gauze CD connected to an electroscope. Ultra-violet light enters through the quartz plate EF, and then a magnetic field is applied.

To this end he employed the usual arrangement of a small negatively charged zinc plate on which ultra-violet light from a distant arc-lamp could shine; the light passing through a plate of quartz, and also through a parallel piece of wire gauze connected with an electrometer. The distance between the zinc plate and the metallic gauze was variable, and the experiment consisted in observing how much electricity reached the gauze from the negatively charged plate, under the influence of light; first without, and then with, a magnetic field of measured strength applied across the region between them.

A little calculation of extreme beauty showed him that the paths of the flying particles under magnetic

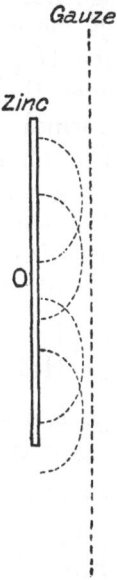

Fig. 9.—Diagram showing the theoretical paths of the electrons emitted from clean negatively-charged zinc in ultra-violet light, under the influence of a strong perpendicular magnetic field; the zinc and gauze being a magnified representation of the AB and CD in fig. 8. The rays would naturally reach the gauze and convey a current to the electrometer, but under the influence of the magnetic field their paths become cycloids and they fail to reach the gauze unless it is brought nearer. The critical distance between the zinc and the gauze—when they are just able to reach it—is what is measured.

influence would be *cycloids*, whose generating circles contained the ratio m/e as well as the ratio E/H^2; that is to say their trajectory, if it could be observed, would involve the electrochemical equivalent required, and likewise the ratio of the electric to the magnetic field applied, as well as the absolute strength of the magnetic field.

The calculation is so simple that it may be given here :

Let figure 9 show the zinc and the gauze facing each other, close together, with a gradient of potential $(V - V')/d = E$ between them ; and let a magnetic field of density H (which letter is, in this case, to represent μ times the intensity, for brevity of writing) be applied normal to the paper.

Then the motion of a charged particle detached and propelled from the origin into the region between the plates, provided that the plates are in vacuum so that there is no resisting medium to interfere, will be—taking the axis of x as a perpendicular to the plate—

$$m\ddot{x} = Ee - He\dot{y}$$
$$m\ddot{y} = He\dot{x}$$

the initial values of x, y, $\dot{x}$, and $\dot{y}$, being all zero.

The solution of these equations, under these initial conditions, is

$$x = a(1 - \cos bt)$$
$$y = a(bt - \sin bt)$$

where $a = \dfrac{Em}{H^2 e}$ and $b = H \cdot \dfrac{e}{m}$;

and from these we see that x is oscillatory in accordance with a versine, ranging from 0 to $2a$ and back ; while y is both oscillatory and progressive, completing its period in a time $\dfrac{2\pi}{b}$, and increasing in every such period by the amount $2\pi a$. In other words, the equations represent a cycloid traced by the rim of a circle of radius a rolling on the zinc plate.

There is no known way of actually observing this quite invisible and purely theoretical trajectory; but when it is perceived that in accordance with this theory all the particles moving between the plates will have similar paths—so far as they do not come near the edge of either plate, in which case they would not be propelled so far—it becomes plain that there should be a critical distance, within which the gauze would receive and intercept *all* the particles, and beyond which not a single one would be able to reach it.   In the figure the gauze is depicted as set just beyond the critical distance, so that it would receive no electricity, even though the ultra-violet light were fully shining; but so that if either its distance from the zinc were diminished, or the electric field strengthened, or the magnetic field weakened, the gauze would at once come within range and receive a plentiful supply of charge from the hypothetical cycloidally-flying particles.   And the critical distance at which this would happen—a thing easily experimentally observed—would be independent of the brightness of the ultra-violet light, and would merely equal the diameter of the generating circle.   In other words,—the critical distance between the plates, when effective transfer of charge occurred, should be $2a$, or $\dfrac{2mE}{eH^2}$; a quantity which by this ingenious means is therefore measurable.   Wherefore the ratio m/e for this case can be experimentally determined, if E and H are both known.   The apparatus employed was shown in Fig. 8.

The sharpness of actual experimental observation of the critical distance was not found quite so great as this simple theory would indicate, because of disturbing causes; one of which was the presence of some residual air, interfering with the perfectly free path of the moving bodies. Nevertheless it was sharp enough for fair determination; and the result was again, in this case also, that the ratio e/m came out 10^7 c.g.s., or more exactly 7×10^6; corresponding closely with the values found by J. J. Thomson, confirmed subsequently by both Lenard and Kaufmann, for the cathode ray particles.

Another phenomenon on which measurements were made was the discharge of negative electricity from an incandescent carbon filament in an atmosphere of rarified hydrogen. This also is subject to disturbance by a magnetic field, as was shown by Elster and Geitel; and a series of measurements, on lines similar to the preceding, resulted in a value

$$\frac{e}{m} = 8 \cdot 7 \times 10^6 \text{ c.g.s.,}$$

a value of the same order of magnitude as before: one thousand times greater than the electrochemical or electrolytic value for hydrogen, and many thousand times greater than for other substances, but always constant and independent of the nature of the substance present.

Another method for measuring these quantities, in what may seem a more direct manner, was devised by Professor Lenard and is depicted in fig. 10. It will be realised that the remarkable character of these experiments is the fact that nothing is visible: there are no cathode rays to be seen, nor any phosphorescent spot produced; all that can be observed is either a maximum deflexion of the electrometer, as by Lenard's plan, or a zero deflexion suddenly changed to a finite deflexion, in J. J. Thomson's plan.

Lenard's method may be described as follows:—
Light from the source L, which is a spark between zinc electrodes, passes through a quartz plate Q, where it enters the vacuum and falls upon a negatively electrified aluminium plate C. B is a perforated earthed screen through which the particles are shot till they fall upon the plate E, which is connected to an electrometer; or if deflected by a magnet they will fall upon the plate F. The amount

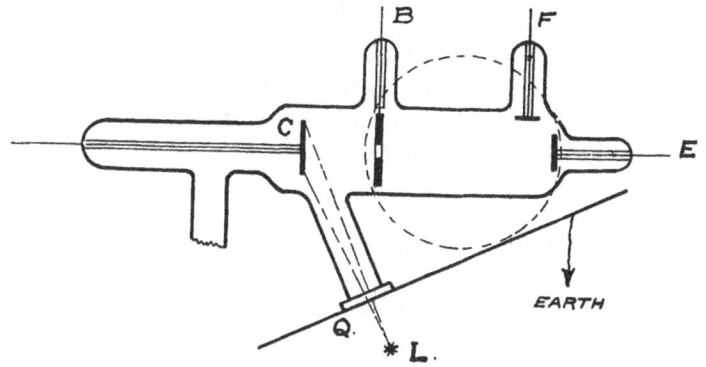

FIG. 10.—Lenard's apparatus for measuring the electrochemical equivalent of the discharge of negative electricity from a cathode illuminated by ultra-violet light in vacuo.

of charge received by these two terminals is separately measured and plotted, with magnetic field as abscissa. The charge received by E will decrease, and that received by F will increase, as the magnetic field is increased, until a maximum is received by F; which will happen when the centre of the plate is the middle of the stream of rays. A further increase in the magnetic force will cause the charge received to fall off. The field required to give this maximum is measured—being determined by metrical examination of the plotted curve. The paths of the particles between B and F must be circles,—since they are not subjected to any but a deflecting force when they have passed through the screen. When the maximum is received by the plate F, the tangent to the mean circle will be horizontal at the position of the hole B; and this fact, together with the passing of this circular trajectory through the centre of the plate F, is sufficient to determine it, and so to give its radius of curvature,—which will be equal to

$$\frac{mu}{e\mathrm{H}}.$$

The velocity u is separately estimated by assuming that it is acquired under electrical influence between C and B, the equation being

$$\tfrac{1}{2}mu^2 = e\left(\mathrm{V} - \mathrm{V}'\right),$$

the latter factor being the difference of potential between C and B.

Thus the two quantities u and e/m are determined.

Positive and Negative Carriers.

In connexion with the above experiments it is important to notice that the above value of e/m for the negative carriers is only obtained *at low pressures*. If the pressure is high the ordinary electrolytic value of e/m, or something still smaller, can be obtained; and this suggests that at ordinary pressures the electron becomes loaded up by attaching itself to an atom, or even by collecting round itself a group of atoms. Further evidence of this is afforded by the fact that there is but little difference between the velocities of the positive and negative ions, when urged by an electric field, in a gas at atmospheric pressure. It will be seen, further on, that in the case of the electrons discharged from Radium, their velocity is so great that this loading effect is imperceptible.

Nothing larger than the ordinary electrolytic value for the e/m ratio is ever given by the *positive* carriers. These are not so easy to observe, but WIEN[*] has examined them by detecting and measuring the slight magnetic deflexion exhibited by certain rays behind a perforated cathode in a vacuum tube, which Goldstein discovered and called *Kanal-strahlen*, and which Wien and Ewers proved were carriers of positive electricity. Wien has shown that they move fairly quickly—that is to say about 360 kilometres per second—in spite of the fact that in hydrogen their ratio e/m is of the order 10^4, that is to say the proper value for a hydrogen atom or ion. With other substances the ratio has been found to vary with the substance and approximately to equal the electrolytic value, for these positively charged atoms. J. J. Thomson has likewise made measurements on the positive carriers, by means of the discharge from incandescent filaments and other positively charged hot bodies, and has confirmed Wien's results—obtaining an electrolytic value for their electrochemical equivalent.

Thus it is forcibly suggested that whereas the positive carriers of electricity are always *ions*, consisting of a unit + charge associated with an atom, the negative carriers are sometimes dissociated from the main bulk of the atom, as if they were only fractions or fragments or constituents or appendages of an atom. These, detached and flying loose, are able to attain to prodigious speed. For any acceleration to which they are subjected is a thousand-fold

[*] Wied. *Ann.* lxv. p. 440. See also Ewers in Wied. *Ann.* lxix. p. 187.

greater than it is even for an atom of hydrogen,—
weighed down and burdened as that is with a mass
of inert material, and subject only to the very same
propulsive force.

Think of the mobility of a particle which experienced
the usual gravitation pull and had only $\frac{1}{1000}$ of the
corresponding mass to carry. Such a mobile particle
as that would drop under the influence of gravity, not
16 feet in the first second, as everything we know
does near the surface of the earth, but 16,000 feet,
or about three miles; and would in one second
acquire under gravity a velocity of six miles per
second; enough almost to carry it out of the range
of the earth's attraction altogether, and more than
enough to carry it round the world, if fired horizon-
tally with such a speed.

The acceleration to which particles are subject in a
vacuum tube is far greater even than this, because
there the forces are so prodigious; gravitation force
on ions is almost infinitesimal compared with common
electrical force on their charges. Suppose, for instance, that they are in a field such as may easily occur
in a vacuum tube, of 3,000 volts per centimetre, one-
tenth of what ordinary air will stand, or ten electrostatic units. The force urging one of these carriers
to move is then $10 \times 10^{-10} = 10^{-9}$ dyne; the mass
being moved, if it is a whole atom of hydrogen, e.g.
if it were a positive carrier in a hydrogen atmosphere,
is only 10^{-24} gramme; and accordingly the acceleration
it experiences is 10^{15} centimetres per second per second,
or a billion times g. Whereas if it were the smallest
kind of negative carrier, its acceleration would be a
thousand times greater still.

The velocity acquired in passing over a distance of
five centimetres under this force is obtained by finding

the square root of $2fh$; that is to say, it is 10^8 centimetres per second for a positive carrier, and 3×10^9
centimetres per second for a negative carrier; and
these are approximately the orders of magnitude
actually observed.

Thus the hypothesis becomes more and more justified
that units of positive charge are always associated
with atoms, in operations which we can control,
and are consequently always complete ions; while
the units of negative charge appear in some cases
with a separate existence,—perhaps carrying with
them part of the atom, in which case they might be
called corpuscles, having a material nucleus; perhaps
pure disembodied electricity, whatever that may
be—an electrical charge detached from matter,—a
mere complexity in the ether, in which case they
would correspond with those hypothetical entities
familiar in theoretical and mathematical treatment as
" electrons."

CHAPTER VII.

IONISATION OF GASES.

IT is constantly necessary to speak of the air or other medium being 'ionised' by the passage of rays, and by many other processes. The term means that the molecules are split up, or dissociated, into their constituent atoms; which, being oppositely charged, form anions and cations respectively. It appears to be an effect due to the violent encounter of an energetically flying particle with a comparatively stationary molecule : a sudden electric wave or pulse, if thin and forcible enough, may do the same; the effect in any case is to break the molecule asunder into constituent ions, some positively, some negatively, charged, thereby converting the medium into a true gaseous electrolyte, until the dissociated atoms have had time to recombine—a time which may be measured in minutes. It is owing to this effect that X-rays are able so readily to discharge an electroscope or other charged body, whatever the sign of its charge may be ; for it is manifest that any ions in the atmosphere of opposite sign will be attracted to it and will neutralise any charge it may possess. Ionised air can always be detected by its making electroscopes leak, irrespective of any defect of insulation in solid supports.

The ionising power of X-rays was observed by Righi soon after their discovery, and almost simultaneously by other physicists.

The terrestrial atmosphere out of doors is usually more or less ionised, and in underground air the ionisation is still more marked, probably owing to radio-activity in the soil. Elster and Geitel have

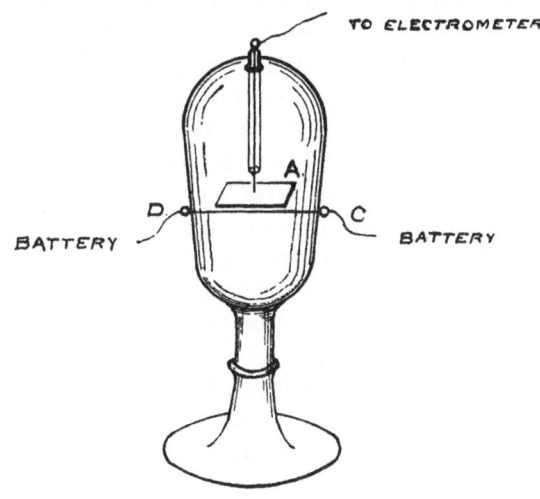

FIG. 11.

examined this matter thoroughly, and Lenard found that even the splashing of water introduced ions into air, so that the atmosphere at base of a waterfall was usually in some degree ionised.

Behaviour of Hot Metals in Gases.

Fig. 11 must stand as representative and typical of a mass of important work on the ionisation and disintegration of material effected by high temperature; it represents the simple apparatus of Elster

and Geitel by which they examined in great detail
a number of incandescent metals in the form of a
wire or strip heated by an electric current in different
gases. In J. J. Thomson's variety of experiment
this current was produced in the secondary of a
transformer, so as to be conveniently insulated. The
details are long and complicated, and must be referred
to in J. J. Thomson's book on *The Conduction of
Electricity through Gases*; but briefly it may be
said that when the gas in the vessel is air, the metal
plate A receives a positive charge when the wire is
heated to a dull red glow; as it becomes hotter the
charge increases until the wire is at a yellow heat;
if it is made hotter than that, the charge diminishes,
and, at the highest temperature, is very small. In
hydrogen the plate is found to receive negative electricity when the wire is hot enough, though at a lower
temperature it receives positive. To get rid of
occluded gases, J. J. Thomson often kept it red hot for
a week or more. McClelland, Branly, H. A. Wilson,
and others have investigated this phenomenon, which
has also been studied in a different form by Preece
and Fleming—following up a curious observation
made by Edison in connexion with incandescent
lamps. Recent measurements by O. W. Richardson
have done much to reduce it to a definite physical
specification. The evidence that an incandescent
platinum wire disintegrates, or at any rate gives off
material, is the fact discovered by Aitken, that a
cloud becomes visible in a luminous beam, if the
moist gas surrounding the wire be suddenly cooled.
This indeed can occur at a much lower temperature
than incandescence. Mr. Owen (*Phil. Mag.*, Sept.,
1903), found that when a platinum wire was in air
there was always a cloud when the temperature of

the wire was raised to about 300° C., even after
long-continued previous incandescence of the wire ;
whereas in pure hydrogen the wire had to be raised
to red heat before a cloud formed.

Measurements of Ionisation Current.

The whole subject of the dissociation or breaking up
of molecules which is effected in gases by Röntgen rays,
by radium radiation, and by a great number of other
influences—a process which is known as ionisation,
because the broken constituents of the molecule are
oppositely charged—is too large to be conveniently
entered upon here. It has been worked at by a
multitude of experimenters, and for an account of
their results the work of J. J. Thomson on *The
Discharge of Electricity through Gases* must be
referred to. It may suffice to say that the products
of decomposition, though in the first instance no
doubt simple ions, seem to be speedily complicated
by the aggregation of other molecules round them ;
and accordingly the diffusion or progress of these ions
is liable to be retarded, and, when measured, is found
slower than might otherwise have been expected.
On the whole, the negative ions tend to move faster
than the positive ones, but the difference is not
necessarily greater than can be observed in liquid
electrolysis. Recent work by H. A. Wilson and
E. Gold, on conduction in flames, has however shown
that the negative carriers in that case are free
electrons.

One of the easiest things to measure is the conductivity of air in this ionised condition, that is to
say, the total current transmitted by it between two
electrodes immersed in it and connected to some
sufficiently sensitive current-measuring device—a

galvanometer in some cases, or an electrometer arranged so as to show a measured leak in others.

A saturation value of the current can thus be found, when all the ions present are taking part in the action to the full extent; the current then reaches a maximum, which cannot be exceeded; and its amount would furnish an estimate of the number of ions present, if the speed and the charge of each were known. Measurements of the total current Neu, the quantity of electricity conveyed per second, have been made by Lenard[1] and Righi[2] and Thomson,[3] and in various gases by Rutherford,[4] now Professor at Montreal; by Beattie[5] and de Smolan at Glasgow, by Zeleny[6] of Minnesota, by McClelland[7] on hot gases from flames, by McLennan[8] of Toronto, by Richardson,[9] H. A. Wilson,[10] and Owen[11] on incandescent filaments. Townsend also has made many experiments on the diffusion speed of ions.

Professor Zeleny measured the velocity by a safe and direct method of making the particles fly down a tube against a wind, and observing the rate of the current of air which was just able to withstand their progress : these measurements constituting a satisfactory confirmation of Thomson's and Rutherford's more indirectly inferred results.

---

[1] Wied. *Ann.*, vol. 63, p. 253.

[2] *Rend. della R. Accad. dei Lincei*, May, 1896.

[3] *Phil. Mag.*, November, 1896.

[4] *Ibid.*, November, 1896, and April, 1897.

[5] *Ibid.*, June, 1897.   [6] *Ibid.*, July, 1898.   [7] *Ibid.*, July, 1898.

[8] *Phil. Trans.*, vol. 195, p. 49, 1899.

[9] *Ibid.*, vol. 201, 1903.   [10] *Ibid.*, vol. 202, 1903.

[11] *Phil. Mag.*, August, 1904.

Condensation of Moisture Experiments.

C. T. R. Wilson investigated the amount of expansion required by both positive and negative ions to act effectively as nuclei in condensing moisture. The ions were produced by Röntgen rays, and electrolytic terminals were inserted to effect a separation of the ions. He found that with an expansion such that the ratio of the final to the initial volume was 1·25, there was a fog in the half of the vessel which contained negative ions, and hardly any condensation in the half containing positive ones, but that when the expansion-ratio was as much as 1·31 there was little or no difference to be seen between the two halves. Thus proving that the negative ions are more efficient, as centres of condensation for water-vapour, than the positive ions.

It may be doubted whether electrons themselves are able to act as nuclei and condense vapour round them, and it appears unlikely that they can do that without the aid of one or more atoms of matter. On this subject Mr. C. T. R. Wilson has favoured me with the following opinion :—

" I certainly think that in practically all cases the electron has already been loaded to form a negative ion, before the expansion by which the necessary supersaturation is brought about has been effected. In air which contains less moisture than corresponds to a four-fold supersaturation, equilibrium is probably reached when the electron has collected around itself a group of molecules not much larger than the ion in dry air. If however a four-fold supersaturation is exceeded, the conditions become unstable and the cluster of molecules increases to a visible drop. If electrons enter the expansion chamber immediately after an expansion, while the four-fold supersaturation is still exceeded, no doubt they will form nuclei for the condensation of drops without any pause at the ion stage. But in the ultraviolet-light experiments I have no doubt that the electrons, or all but an insignificant proportion of them, had been loaded up and had existed as ions before the expansion was made."

The next chapter contains an account of what are probably the most important experiments yet made in the Cavendish Laboratory.

CHAPTER VIII.

DETERMINATION OF THE MASS OF AN ELECTRON.

So far, all the measurements quoted have resulted in a consensus of certainty respecting our knowledge of e/m for gaseous conduction and radiation; and the measurements made on the cathode rays in a Crookes's tube, or near a plate leaking in ultra-violet light, have likewise given us a knowledge of their velocity, and shown that it is about one-thirtieth of the velocity of light: more or less according to circumstances. But so far no direct estimate has been made of either e or m separately. The difficulty of making these measurements is great, because we are dealing with an aggregate of an enormous and unknown number of these bodies. It would not be difficult to make a determination of the aggregate mass of a set of projectiles, say $N m$, where N is the number falling on a target in a given time, by means of the heat which the blow generates; or better, perhaps, by the momentum which they would impart to a moving arm after the fashion of a ballistic pendulum; provided their velocity u were known, as in this case it is. The aggregate energy $\frac{1}{2} N m u^2$, or the aggregate momentum $N m u$, could thus be found; but how is m to be separated from N?

Again, if the particles are collected in a hollow

vessel attached to an electrometer of known capacity,
it is not difficult to estimate the total quantity of
electricity which enters the vessel in a given time,
that is to say, to determine Ne; but, again, how are
we to discriminate e from N?

Another thing that is comparatively easy to determine, especially in such cases as leak from a negative
surface under the action of ultra-violet light, or the
conductivity of air induced by the influence of Röntgen
rays, is the total current transmitted; viz. the quantity
Neu, the quantity of electricity conveyed per second
between electrodes immersed in the air, and maintained at a sufficiently high difference of potential
to cause all the corpuscles or the ions present to
take part in conducting the current.

We may consider the following quantities experimentally determined, by researches carried on at
the Cavendish laboratory and elsewhere, and so far
already described or indicated in the preceding three
chapters :—

$$e/m$$
$$u$$
$$Ne$$
$$Nm$$
$$Neu$$

But still we have not described a method of
measuring separately either e or m: only methods
of measuring their ratio.

If only it were now possible to *count* the corpuscles
or electrons,—to determine the number N which are
started into existence, or which enter the hollow
vessel or which take part in conveying the current in
the case of a leak by ultra-violet light,—we should no
longer have to *guess* at the actual value of e and of m
separately, but should have really *determined* them.

This brilliant research has actually been carried out by Professor J. J. Thomson, by means of a method partly due to Mr. C. T. R. Wilson, supplementing a fact discovered by Mr. Aitken, and interpreted in the light of hydrodynamic principles laid down long ago by Sir George Stokes.

I must be excused for waxing somewhat enthusiastic over this matter : it seems to me one of the most brilliant things that has recently been done in experimental physics. Indeed I should not need much urging to cancel the "recently" from this sentence ; save that it is never safe for a contemporary to usurp the function of a future historian of science, who can regard matters from a proper perspective.

The matter is rather long to explain from the beginning, and I must take it in sections.

Aitken and Cloud Nuclei.

First of all, Mr. John Aitken,* of Edinburgh, discovered in 1880 that cloud or mist globules could not form without solid nuclei, so that in perfectly clear and dust-free air aqueous vapour did not condense, and mist did not form. (See, for instance, my lecture to the British Association at Montreal, in 1884, on "Dust"—*Nature*, vol. 31, p. 268.)

Without solid surfaces, in clear space, vapours could become supersaturated; but the introduction of a nucleus would immediately start condensation, and according to the number of nuclei, or condensation centres, so will be the number of cloud globules formed.

Every cloud or mist globule is essentially a minute falling raindrop, not floating in the least, but falling through a resisting medium—falling slowly

* *Trans. R.S. Edin.*, vol. 30, pp. 337-368 (1883).

because it is of such insignificant weight compared with its surface—but falling always relatively to the air. A cloud may readily be carried up by a current of air, but that is only because the air is moving up faster than the drops are trickling down through it. No motion of the air disturbs the *relative* falling motion : the absolute motion with reference to the earth's surface is the resultant of the two.

The fact that nuclei are required for mist precipitation can be proved by filtering them out with cotton wool, and finding that as the nuclei get fewer the mist condensation differs in character, becoming ultimately what is called a Scotch mist, such as forms in fairly clean air ; where since the dust particles are comparatively few, the centres of condensation are few also, and accordingly have each to condense a considerable amount of vapour; so that the drops are not nearly so close together and are bigger; wherefore they fall quicker, like very fine rain. In perfectly clean elaborately-filtered air the dew point may be far passed without any vapour condensing, and the space will remain quite transparent in spite of its being supersaturated with vapour.

The reason for this effect of, and necessity for, nuclei, is thrown into strong relief by Lord Kelvin's theory concerning the effect of curvature on vapour tension,* because the more a liquid surface is curved the more it tends to evaporate, and an infinitely convex surface would immediately flash off into vapour. Consequently an infinitesimal globule of liquid cannot exist ; vapour can only condense on a surface of finite curvature, such as is afforded by a dust particle or other body consisting of a large aggregate of atoms. For it must be remembered

* See, for instance, Maxwell's *Theory of Heat*, 1891 edition, p. 290.

that a single grain of lycopodium powder contains about a trillion atoms, and a dust particle big enough to condense vapour need not consist of more than a billion, or perhaps not more than a million, atoms, and need by no means be big enough to be visible even in a microscope. It is, however, material enough to be stopped by a properly packed cotton-wool filter.

J. J. Thomson and Electrical Nuclei.

In 1888 it was shown by J. J. Thomson, in his book *Applications of Dynamics to Physics and Chemistry*, p. 164, that electrification of a body would partially neutralise the effect of curvature, and so assist the condensation of vapour on a convex surface.

Consider a drop of liquid, or a soap bubble; the effect of the curvature of the surface is to give a radial component of surface tension inwards, causing an increased pressure internally. The effect of electrification is just the opposite: it causes a direct pressure outwards, which goes by the name of electric tension.

The way these depend on size is as follows:

The radial-pressure component of the surface-tension T is

$$\frac{2T}{r} \text{ inwards.}$$

The electric tension is

$$2\pi \overline{\kappa} \overline{t}^2 = \frac{e^2}{8\pi \kappa r^4} \text{ outwards.}$$

They are differently affected, therefore, by the size of the globule; hence at some size or other they must balance, and such an electrified convex surface will

behave as if it were unelectrified but flat. Accordingly vapour which would refuse to condense on an
unelectrified convex surface, until far below the dew
point, will begin to condense on it, if sufficiently
electrified, the instant the dew point is reached.

The critical size at which the ionic charge enables
a sphere of water to act as regards condensation as if
it were flat, can be reckoned by equating the pressure
to the tension, thus :

$$\frac{2T}{r} = \frac{e^2}{8\pi \kappa r^4}$$

$$\text{or} \quad r^3 = \frac{e^2}{16\pi \kappa T} = \frac{10^{-21}}{50 \times 80} = \frac{1}{4} \times 10^{-24} \text{ c.c.}$$

whence $r = 10^{-8}$ approximately, or is of atomic
magnitude.

Hence *ions* may be expected to condense vapour
at the ordinary dew-point; and anything bigger
which possesses the same charge can condense it
still more easily.

Accordingly an electric charge assists vapour to
condense; and a sufficient electric charge might
cause it to condense on quite a small body—as
small even as an atom. Hence in the presence of
ions, dust particles are not necessary for condensation. Vapour may condense on these electrical
nuclei without the need for solids of finite curvature. The electrical nuclei cannot be completely
filtered out by cotton wool: they will exist or
can be produced in dust-free air. No doubt if
they are passed through a great amount of metal
gauze they may be diminished in number, but
they are not easily got rid of except by their
own diffusion, which does ultimately enable them
to pair off or to migrate to the sides of the vessel.

They can be got rid of most quickly, however, by introducing an electric field, that is to say by supplying electrodes maintained at a few volts difference of potential. They will then immediately make a procession, as in electrolysis, only with much greater speed, because their motion is much less resisted or interfered with by chance collisions; so they will soon reach and cling to their respective electrodes, and in that case again no true mist can form. But it must be remembered that any of the numerous causes of ionisation can produce some of them again.

While ions are present in considerable numbers a thick mist will form whenever the space is saturated with vapour, but it will be a mist of different appearance from the slight rain-like condensation which may be seen forming round the few residual dust particles. The mist globules will usually be of uniform size, and some estimate of that size can be roughly attempted by the diffraction colours which can be seen if a point of light is looked at through the mist: not, however, a very easy plan for making a trustworthy estimate.*

Electrical nuclei can be produced in various ways—by anything, in fact, which dissociates the air or which fills it with ions. Some are produced by the splashing and spray of water; some are given off from flames, and from red-hot bodies; they are produced in considerable numbers when Röntgen and when Becquerel rays travel through air; they can be given off by radio-active substances like uranium; and they are easily emitted by a negatively charged metallic surface exposed to ultra-violet light.

* See C. T. R. Wilson, *Phil. Trans.*, 1897, A, vol. 189, p. 283.

Wilson and Metrical Cloud Condensation.

Mr. C. T. R. Wilson,* in his study at the Cavendish Laboratory of cloud formation under the influence of Röntgen rays and other agents, devised a plan for precipitating a definite and known quantity of aqueous vapour in a visible form. This was done by an arrangement for making a sudden or adiabatic expansion of saturated air, and making it to a carefully measured amount. The apparatus employed is shown in Fig. 12.

One test-tube moving inside another is employed as a piston, and by a certain arrangement the piston was enabled to drop with great suddenness and thus to produce a measured small exhaustion and consequent cooling in the reservoir containing the gas under experiment; saturated as it is with vapour, and supplied with electric nuclei. The mist at once formed, and the drops began to fall slowly, as usual. Mr. Wilson tried to get an estimate of their size from the colours, but it was difficult and unsatisfactory. If the size had been known, their number would have been known too, because the measured amount of expansion had produced a known fall of temperature below the dew point, and so had condensed a known amount of aqueous vapour, which would be distributed equally among all the equal globules.

It occurred to J. J. Thomson that a better estimate of size could be made by observing their rate of falling, which is a thing not difficult to observe since they all fall together, being all of the same size. In any mist formed in a bell-jar it is easy to watch it settling down, by watching its fairly definite upper

* *Phil. Trans.*, A, 1897, vol. 189, p. 265.

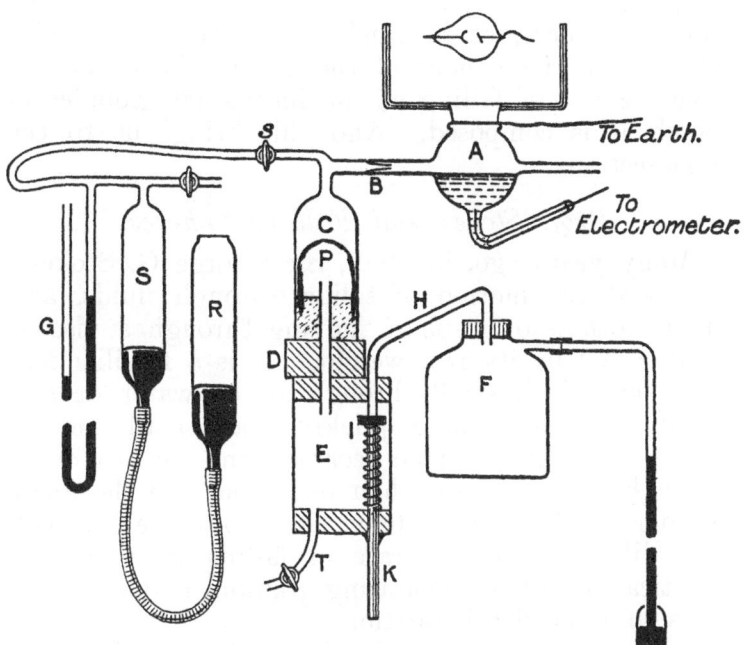

FIG. 12.—A represents one of the vessels in which the fog is formed
whose rate of fall is to be measured by Mr. Wilson's method: it is
arranged for the ionisation produced by X-rays. The vessel A, containing some water, and covered by an earthed aluminium plate, is
in communication with a vessel C through the tube B. Inside C is a thin-
walled test-tube P, which serves as a piston or bell-jar (shown rather
too stumpy) and with its lip always dipping under water like a gasometer.
D is an indiarubber stopper closing the end of tube C: the lower part
of the tube C ought to be shown filled with water to such a height that
the mouth of the piston is always below the surface. A glass tube con-
nects the inside of the test-tube P with a space E. The space E may
be put in connection with an exhausted space F through the tube H.
The end of the tube H, inside the space E, is ground flat, and is closed
by an indiarubber stopper I, which is kept pressed against the tube H
by means of a spiral spring. The stopper I is fixed to a rod K ; by
pulling the rod down smartly the pressure inside the test-tube is
lowered, and the piston P falls rapidly until it strikes against the
indiarubber stopper D. The falling of the piston causes the gas in
A to expand : the tubes R and S are for the purpose of varying the
amount of the expansion. Before an expansion the piston P is raised
by admitting air through T, which is then closed. Then, when everything is ready, K is pulled, and the cloud forms in A.

surface, a clear space being left above it which gradually increases in thickness as the cloud falls. The rate of movement of the top of the cloud will give the rate of falling of the individual globules of which it is composed. And this brings us to the next section.

Prof. Stokes and Falling Spheres.

Many years ago, in 1849, Sir George G. Stokes* discussed the motion of solids through fluids, and among others of a sphere moving through a viscous fluid urged by its own weight. It is a familiar fact that large bodies fall through air or water or any resisting medium more quickly than small ones of the same shape. Thus coarse sand settles down through water quicker than fine sand, and the finest powder takes a very long time to settle; in fact this difference of the rate of falling is used as a practical process of separating granular materials into sizes, and is called levigation.

So it is in air : large raindrops fall violently, small raindrops fall gently, and mist globules hardly fall at all—fall so slowly that their motion is difficult to observe,—but the same law governs all, so long as the motion is not too violent, or so long as the falling body has no edges such as will cause eddies during the fall. A sphere falling slowly, controlled by viscosity alone without waves or eddies, is the simplest case. It soon reaches what is called a terminal velocity—the speed at which the viscous resistance exactly balances its weight. At this speed it is subject to zero resultant force, so it simply obeys the first law of motion and moves at a constant speed.† This constant speed or terminal

* *Camb. Trans. Phil. Soc.*, ix. 48.   † Cf. *Nature*, vol. 31, p. 266.

velocity was calculated by Sir George Stokes for the case of a falling raindrop of radius r as follows :

$$c = \frac{2}{9} \frac{g\rho r^2}{\text{viscosity of air}},$$

where ρ is the excess density of the sphere over the medium it moves in; provided there is no finite slip at the surface. The maximum possible effect of surface slip—which will occur to some extent when the falling globules are very minute—is to make the possible terminal velocity half as great again : in other words to convert the numerical coefficient $\frac{2}{9}$ into $\frac{1}{3}$.

This simple formula gives the connexion between the rate of fall of any small rain or fogdrop and its size ; and by observation of this speed, therefore, knowing the viscosity of air, it is possible to calculate the dimensions of the falling drops.

J. J. Thomson's Experiment of Counting.

We have now all the materials ready for understanding the experiment to be performed,* so as to count the ions which are produced in air under the influence of Röntgen rays, or which have been produced from a negatively electrified surface illuminated with ultra-violet light. The apparatus for the former is depicted in Fig. 12. The ionising beam of X-rays enters the chamber A from above, through an earthed aluminium lid which keeps it airtight.

The rate of leak, which must be observed in order to calculate Neu, is determined by connecting the water and the aluminium plate to the terminals of

* *Phil. Mag.*, December, 1898, and December, 1899.

an electrometer; a sufficient difference of potential being maintained between them.

And now, metrical condensation having been produced by the expansion appliance, and a mist formed, the rate of its fall or gradual subsidence can be observed by looking through the vessel at an illuminated surface; whence by Stokes's theorem the size of each globule is known. The quantity of water which had gone to form globules is known from the measured amount of expansion, by a process the details of which I will not give here; and so the number of such globules, and therefore the number of their condensation-centres or nuclei or ions, can be determined.

If c is the observed rate of fall in stagnant air, the linear dimensions of the falling drops will be

$$r = \sqrt{\left(\frac{9\mu c}{2g\rho}\right)} = \sqrt{\left(\frac{4\cdot5 \times \cdot00018}{981}c\right)} \text{ centimetres.}$$

In a given case c was observed to be 0·14 centimetres per second; hence the volume of each drop was in that case

$$\frac{4}{3}\pi r^3 = 1\cdot6 \times 10^{-10} \text{ c.c. ;}$$

and so, if the aggregate amount of water in all the drops in a given space is reckoned from the measured amount of adiabatic expansion which caused the chill and the precipitation, the drops can be counted.

Professor Thomson, a few months later, repeated this experiment, the ions being now produced by illuminating a negatively electrified zinc plate with ultra-violet light. The apparatus used is shown in fig. 13. A clean zinc surface *in vacuo*, faced by a transparent conductor through which the light could shine on it, and by a window of quartz which makes

the vessel airtight so that it might be exhausted and
yet allow the ultra-violet light to pass, was employed,
and connected to the expansion apparatus, fig. 12,
instead of the vessel A. The current Neu was
measured just as in the original experiment.

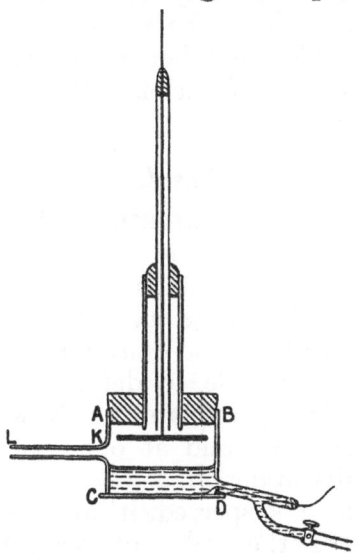

FIG. 13.—J. J. Thomson's auxiliary apparatus for the counting experiment. Ultra-violet light passes through the quartz plate CD and
through a layer of liquid, which keeps the air saturated, and which
constitutes one electrode, to a clean zinc plate K, which constitutes the
other, and which is kept negatively electrified. Connexion through
the tube L is made with the expansion apparatus shown in fig. 12, this
being employed instead of the vessel A in that figure. Then when
the sudden measured expansion is caused, a fog is condensed round the
negative ions which have sprung into being in consequence of the
electrons thrown off by the ultra-violet light; and the rate of settling
down of this cloud is then measured.

A great many precautions must be taken, because
there will be *some* residual cloud found even when
electrons or intended nuclei are not present. Positive
ions and other stray or undesired nuclei—if present—
can be eliminated by aid of their different behaviour.
A differential observation is generally necessary;

moreover care must be taken to ensure that all the desired nuclei are utilised, and not only a portion.

The number of drops found in a certain experiment, by this means, was about 30,000 to the cubic centimetre; the total quantity of water which went to form them being about the two-hundredth part of a milligramme. The number of drops is of course equal to the number of nuclei. Wherefore the nuclei are counted.

Result.

The result of the execution of this ingenious counting process is that the absolute charge and the absolute mass of an electron are at length directly determined. Hitherto we have determined by many and various ways the ratio e/m and the speed u. We have likewise been able to determine Neu and Ne and Nmu^2, as already explained. Now at length we have determined N; and at once the terms in the ratio e/m are disentangled.

e comes out, as suspected, in all cases, the regular ionic charge, of the order of magnitude 3×10^{-10} electrostatic, or 10^{-20} electromagnetic units. Hence, while m comes out for positive carriers and for all ions the appropriate mass of the atoms present,—or in some cases greater than this, by reason of the formation of molecular aggregates,—for the negative carriers set free by ultra-violet light, and for the other cases where $e/m = 10^7$, the masses come out definitely of the order 10^{-27} grammes; or about $\frac{1}{700}$th part of the smallest and lightest previously known quantity of matter, viz., an atom of hydrogen.

The existence of masses smaller than atoms is thus experimentally demonstrated, and a discovery clinched of epoch-making importance.

CHAPTER IX.

FURTHER DETAILS CONCERNING ELECTRONS AND IONS.

Confirmatory Measurements of Charge.

A CONFIRMATORY measure of the charge e, can be made by first observing the rate of fall of a cloud condensed round the ions acting as nuclei—a measurement which gives the weight w of each drop,—and, then applying to the cloud a vertical electric field and adjusting its strength until it is just able to check the fall and hang the drops in air like Mahomet's coffin. For under these conditions of course $w = Ee$, and since everything except e has been measured, e is at once known. This ingenious process was devised by Mr. H. A. Wilson, and is described by him in the *Philosophical Magazine* for April, 1903. In practice the above would be slightly modified and a simple differential method employed. The numerical result at which he arrived was that the value of $e = 3\cdot1 \times 10^{-10}$. The result also had the effect of confirming the applicability of Stokes' law, in the cases where *fall* had been permitted.

This seems to be the value obtained for the electric charge associated with every kind of monad ion, both positive and negative, as well as for the separate electrons. The same value characterises even the molecular aggregates which are often found in conducting air or other gas.

Fig. 14 shows H. A. Wilson's apparatus for suspending a condensed cloud against gravity by means of a known electric field. The magnet M, attracting its keeper, opens connexion with an injector pump through a valve V, and thereby, through the action of the test tube T, produces a measured amount of expansion—in C. T. R. Wilson's manner—as exhibited in fig. 12. The expansion could subsequently be measured by the gauge H.

A cloud is formed in the bell-jar B, and the electric field is applied between the plates attached to the terminals C D, kept electrified by a battery of a great number of cells. The battery, however, was only applied after the cloud had formed, otherwise all the ions would be removed from the vessel by electrolytic action. Röntgen rays were used to produce the ions, but the coil exciting the rays was switched off and stopped, by means of an automatically working switch S, the instant before the valve was pulled by the magnet.

J. J. Thomson has recently repeated his original experiments, and by combining a determination of Neu with an independent velocity measurement, he has made what is now regarded as a standard estimate of e, namely, 3.4×10^{-10} electrostatic units.

H. A. Wilson's confirmation of this by a totally different method—the Mahomet's coffin method—by which he finds 3.1×10^{-10}, has already been mentioned.

The importance of obtaining accuracy in these measurements is that thereby lies our chance of determining m with accuracy—the mass either of an electron, or of an atom of matter.

One of the first direct determinations of the average charge on a gaseous ion was made by ·Professor

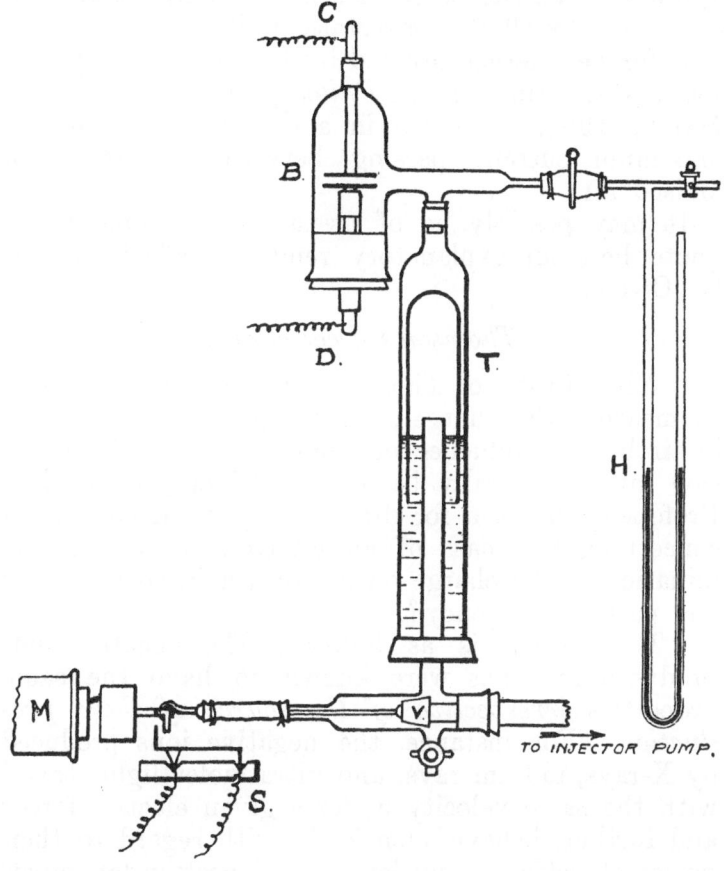

FIG. 14.

Townsend (*Philosophical Magazine*, February, 1898), his method consisting in bubbling newly-prepared gas through water and measuring the charge on each droplet. The deduction must be regarded as rather hypothetical, because it was an assumption that each droplet contained only one ion. Nevertheless, that appears to be the truth, since the charge found on each was 3×10^{-10} electrostatic units.

A further method applied by Professor J. J. Thomson, and described in the *Philosophical Magazine* for March, 1903, was used in a determination of the maximum current passing between electrodes in ionised air.

It may possibly be of assistance to some if we quote here an explanatory remark drafted by Mr. G. Owen :

Thomson's Deductions.

" What Professor Thomson succeeded in doing was to measure the charge on the negative ion produced in air by the influence of Röntgen rays. The question might naturally be asked : What grounds had Professor Thomson for drawing a general conclusion concerning the mass of an electron from a determination of the charge on an ion produced in air by one particular agency ?

" The answer is as follows : The negative ions produced in a gas were known to have the same properties *irrespective of the source of their production*. For instance, the negative ions produced by X-rays, radium rays, and ultra-violet light, travel with the same velocity under a given electric force ; and further, behave identically with regard to their power of acting as nuclei of condensation for supersaturated water-vapour. It was therefore justifiable to assume, in the first place, that the charge on an ion produced by X-rays was equal to the charge on an ion produced by the impact of ultra-violet light on a metal surface. (This conclusion Prof. Thomson verified later on by direct experiment.) Again, the ratio $\frac{e}{m}$ had been determined for the cathode ray particles, and for the carriers of negative electricity from incandescent filaments and from *metal plates illuminated by ultra-violet light* at low pressures, and it will be remembered that in all these cases the *same* value of $\frac{e}{m}$ was obtained, thereby justifying the surmise that the charge on an *electron* was the same as that on an ion produced by ultra-violet light, and therefore the same as on a negative ion produced by X-rays."

Estimate of Size.

On the hypothesis that the flying or vibrating fragment is a material corpuscle charged with electricity, so that it has a duplex constitution and a compound kind of inertia, part material and part electrical, no further progress can be made. But on the hypothesis that the flying or vibrating particle is an electron—a charge of electricity and nothing else—a constituent of an atom but with no material nucleus—so that the whole of atomic properties might possibly turn out to be due to an aggregate of electrons of opposite sign, of which one or two are comparatively free and detachable—on this hypothesis a determination of the mass of a corpuscle carries with it, as a consequence, a determination of its size also.

Because, as has already been pointed out, any required amount of self-induction can be conferred on a wire by making it fine enough, and any required amount of energy can be conferred upon an electric charge by making it concentrated enough. The energy at a given speed of motion will be proportional both to the quantity and the potential, and the latter can be made as great as we please by making the size of the body possessing the charge extremely small.

It is the intense region of force close to the wire or close to the charged particle which is the effective region; and so, as stated, a knowledge of the mass or kinetic energy at a given speed suffices, on a purely electric theory of matter, to determine the size of the electron constituents of which it is hypothetically composed. For whether there be any intrinsically material inertia or not, there certainly is an electrical inertia. The cause of it in the electrical case is known: it is due to the reaction of the electric and magnetic fields during acceleration periods, and is denominated self-induction.

Quite possibly there is no other kind. Quite possibly that which we observe as the inertia of ordinary matter is simply the electric inertia, or self-induction, of an immense number of ionic charges, or electric atoms, or electrons.

This is by far the most interesting hypothesis, because it enables us to progress, and is definite. The admixture of properties—partly explained, viz. the electrical, partly unexplained, viz. the material— lands us nowhere; unless, by some only partially imagined means, we were able to estimate how much of the total appertains to each ingredient.

The mass of a corpuscle has been measured as something akin to $\frac{1}{1000}$ of an atom of hydrogen, and its charge as 10^{-10} electrostatic unit. Now this amount of electricity will have that amount of inertia if it exists on a sphere of radius 10^{-13} centimetre, but not otherwise. Consequently we may assume the size of the supposed pure electron to be of the order 10^{-13} centimetre in diameter; or $\frac{1}{100,000}$ of the linear dimension known as molecular magnitude, viz. 10^{-8} centimetre.

The calculation of order of magnitude is quite simple, for all ordinary speeds; because, for them

$$m = \mu e^2/a,$$

$$\therefore\ a = \frac{e}{m}.\mu e = 10^7 \times 10^{-20} = 10^{-13}\ \text{centimetre,}$$

though it might with some data be estimated as small as 10^{-14}.* Minuteness like that easily explains the penetrating power of cathode rays. Especially if the atoms of matter are themselves composed of such minute particles. For the interspaces will be enormous compared with the filled-up space, and a point can penetrate far into such an assemblage without striking anything.

Penetrability of Matter by Electrons.

The mean free path of a particle is a question of probability. In a space containing n_1 obstacles to the unit volume, a space Ax will contain $n = Axn_1$ of them; and the chances of a collision, while one of them travels a length x, will be approximately their combined areas, as targets, compared with the total area available for both hit or miss—that is to say,

$$\frac{n\pi a^2}{A};\ \text{which we may write}\ \beta x\ \text{or}\ \frac{x}{\bar{x}},$$

where $\bar{x}$ is the "mean free path," or average distance travelled by any one particle without a collision with another, and β the number of encounters while

---

* See Lodge in the *Electrician* for March 12, 1897, vol. 38, page 644, where the size is deduced from the then just discovered Zeeman effect.

travelling unit distance.   But in saying this we are
ignoring the forces between the particles, as well as
their motion, and are not considering a swing round
as a collision.

Nevertheless, as regards order of magnitude—

$$\bar{x} = \frac{Ax}{n\pi a^2} = \frac{1}{n_1\pi a^2} = \frac{d^3}{\pi a^2},$$

where d^3 is the cubic space allotted to each particle,
while $\frac{4}{3}\pi a^3$ is the actual bulk of each.

Therefore approximately

$$\frac{\bar{x}}{a} = \frac{\text{total space occupied}}{\text{a few times the aggregate volume of the particles}},$$

a statement roughly analogous to Clausius' or
Loschmidt's theorem in the kinetic theory of gases.

Hence the mean free path can be estimated by
considering how much space the substance of all the
electrons in an atom occupies, as compared with all
the space which the atom occupies itself.   In other
words, we have to consider what the size 10^{-13} for an
electron's diameter means, as compared with the size
10^{-8} for an atom's diameter.   In the solar system the
diameter of the earth is $\frac{1}{24000}$th part of the diameter
of its orbit round the sun.   Consequently if the earth
represented an electron, an atom would occupy a
sphere with the sun as centre and four times the
distance of the earth as radius.

In other words, if an average atom is composed of
electrons, they are about as far apart in that atom in
proportion to their size as the planets in the solar
system are in proportion to their size.

In an atom of hydrogen there would have to be
roughly 1,000, or say more exactly 700, electrons in
order to make up the proper mass.

In an atom of sodium, which is twenty-three times as heavy, there must be about 15,000 electrons.

And in an atom of mercury there must be over 100,000 electrons, if atomic mass be wholly due to them.

Consider then an atom of mercury containing 100,000 of these bodies packed in a sphere 10^{-8} centimetre in diameter. One would think at first they must be crowded; but there is plenty of room. Each electron is only 10^{-13} centimetre across, and there are only about fifty of them in a row along any diameter of the atom, whereas there might be a hundred thousand in the same length; hence the empty space inside the atom is enormously greater than the filled spaces. At least a thousand times greater in linear dimension, or a thousand million times greater in bulk.

The whole volume of the atom is 10^{-24} c.c.; the aggregate volume of all the electrons composing the atom is $10^{5} \times 10^{-39} = 10^{-34}$ c.c.; consequently the space left empty is 10^{10} or ten thousand million times the filled space.

Even inside an atom of mercury, therefore, the amount of crowding is fairly analogous to that of the planets in the solar system. For though the outer planets are spaced further apart than the inner ones, they are also bigger, to practically a compensating extent.

Now, going back to what is sometimes called Loschmidt's theorem in the kinetic theory of gases, obtained roughly above—

$$\frac{\text{mean free path}}{\frac{1}{8}\text{ diameter of particle}}$$

$$= \frac{\text{volume of space available to particles}}{\text{combined volume of all their substance}}$$

—we have reckoned the latter fraction, in the inside of an atom of mercury, as—

$$\frac{\frac{4}{3}\pi \times (10^{-8})^3}{100,000 \times \frac{4}{3}\pi(10^{-13})^3} = \frac{10^{-24}}{10^5 \times 10^{-39}} = 10^{10}.$$

Hence the mean free path of an *electron* inside an atom of mercury will be comparable to 10^9 times the size of an electron, *i.e.*, it will be 10^{-4} centimetre; that is, it may get through, on the average, the substance of some 10,000 mercury atoms in a row, before collision with anything.

In any other less dense substance it will go further. In ordinary air, on an average free journey, it would escape collision with a hundred million molecules in a room, which would be equivalent to a distance of about four inches. In the case of corpuscles plunging into a dense metal, the actual *distance* achieved by them is very small, only the thousandth part of a millimetre on the average, and it need by no means necessarily be a straight line; so a target of platinum succeeds in stopping them very near its surface, and enables the X-rays generated by the shock fairly to emerge. Some corpuscles will be stopped more suddenly than this, and some will travel further, but 10^{-4} centimetre, or the thousandth of a millimetre, should be comparable with the average distance travelled in a solid as dense as platinum.

Effects of an Encounter.

This distance, however, gives no notion of the value of the negative acceleration during a collision, because the greater part of that thousandth of a millimetre is free flight; the stoppage occurs only as the last episode of that flight, viz. at the instant of collision. The colliding masses are 100,000 to 1, so the change of velocity at impact could be estimated; but the impact will really be more of an astronomical or cometary character, and the effect is analogous to the entrapping of comets when they pass near a planet, thereby rendering them permanent members of the solar system.

The *ordinary* behaviour of a foreign comet, which comes and goes, may be called a collision with, and rebound from, the sun; for although there is no real encounter of main substance, that is what it would appear like if it could be seen from the depths of space; and the two branches of the comet's hyperbolic orbit would look like straight lines of approach and recession.

Comets which happen to pass very near a planet, however, are deflected, swirled round, and often virtually caught by that planet, receding only with an insignificant differential velocity which is unable to carry them away from the attraction of the sun : towards which they then drop. If they do not actually drop into it, they will continue to revolve round it in an elliptic orbit, becoming a member of the solar system, and liable ultimately to be degraded into a swarm of meteors.

This is the sort of process known to occur in astronomy; and circumstances not unlike that may attend the encounter, or apparent collision, of a furiously-flying comet-like electron with part of the massive system of an atom.

The stoppage, therefore, will occur well within the limits of atomic magnitude, 10^{-8} centimetre; and so the acceleration will be of the order $\dfrac{u^2}{2b} = 10^{26}$ c.g.s., and the force needed thus to stop even a single electron will be the tenth of a dyne.

No wonder that violent radiation-effects are produced. The "power" required to stop an electron, flying with one-thirtieth of the speed of light, inside a molecular thickness, can be estimated thus—

$$\text{energy} \div \text{time} = \frac{1}{2} m u^2 \cdot \frac{u}{2b}$$

$$= 10^{-27}(10^9)^3 10^8 = 10^8 \text{ ergs per second};$$

or thus—

$$Fl \div t = \frac{1}{2} Fu = 10^{-1} \times 10^9 = 10^8 \text{ again,}$$

which is equivalent to ten watts. (Though the time it lasts is only the 10^{-17} part of a second.)

But only a small fraction of this goes into radiation. The radiating power can be estimated thus, from Larmor's expression for it, as deduced in Appendix G,

$$\frac{\mu e^2}{v}(\dot{u})^2 = \frac{10^{-40}}{10^{10}} \times 10^{52} = 100 \text{ ergs per second.}$$

The rest therefore, it would appear, must take the form of heat.

It is worth considering what circumstances would give radiation an advantage over heat, and *vice versâ*. Because sometimes conspicuously the target gets heated, and sometimes X-rays are emitted. Let u be the speed and l the distance of stoppage, then

$$\dot{u} = \frac{u^2}{2l},$$

so the force required to stop it is

$$m\dot{u} = \frac{2\mu e^2}{3a} \frac{u^2}{2l}.$$

The "power" of the blow is

$$\tfrac{1}{2}Fu = \frac{\mu e^2 u^3}{6al},$$

whereas the radiation power is—

$$\frac{2\mu e^2}{3v} \cdot \left(\frac{u^2}{2l}\right)^2 = \frac{\mu e^2 u^4}{6vl^2};$$

therefore $\dfrac{\text{radiating power}}{\text{total power}} = \dfrac{a}{l} \cdot \dfrac{u}{v} = \dfrac{2a}{vt}$,

where t is the time of stoppage, and v is the velocity of light.

Hence effective radiating power depends chiefly on very sudden stoppage, and on the speed being near that of light. If the velocity is a tenth that of light, and if an electron can be stopped in something like its own diameter, about 10 per cent. of the energy will go in radiation, and the rest will take other forms, presumably heat. But it would take immense "power" to effect such a stoppage as that: not less than two or three thousand kilowatts for each electron. So probably a stoppage within *atomic* dimension is all that can be expected, and that could be managed by something like 50 watts. But then an exceedingly small fraction of the whole— only about one millionth—would in that case take the form of X-rays; their wave-shell then having a thickness comparable with molecular magnitude: whereas in the previous case it was incomparably thinner, and therefore far more penetrating. Both thicknesses however are very small compared with the wave-lengths of ordinary light.

As the velocity diminishes, more and more of the energy takes the form of heat; which agrees with the fact that at moderate vacua the target gets red-hot.

The ratio of the radiation power to the total power, is as the dimensions of an electron to the distance light would travel during the period of the stoppage : taking the acceleration as uniform. So to get *all* the energy radiated it is necessary to stop a pellet moving with a tenth the speed of light in something like a tenth of its own diameter.

CHAPTER X.

THE ELECTRON THEORY OF CONDUCTION
AND OF RADIATION.

MEANWHILE the probability of the existence of electrons and the possibility of regarding them as the basis of all electric and of most other material phenomena, had seized hold of the imagination of several mathematical physicists, notably of Professor H. A. Lorentz and of Dr. J. Larmor. The former, who was first in the field (1892 and 1895), was driven in this direction by the problem of the astronomical aberration of light and the optics of moving media treated from the electric standpoint. The latter reached the same goal independently, from the dynamics of the free ether, on the basis of MacCullagh and Kelvin, which required discrete mobile sources of disturbance (electrons) as a basis for development. Both these philosophers endeavoured to trace all electric properties to the behaviour of electrons, usually of course in association with material atoms; while Larmor's procedure also impelled him to make intelligible by conceptual 'models' the dynamical possibility of a structure in the ether which should have the properties of an electron, whether positive or negative,—the two being treated as mirror images of each other—and so to reduce a great part of Physics to its simplest terms. This fine attempt, made in 1894, involved definite illustrative conceptions—of the structure of an electron, of its size on the theory that inertia is entirely electric, of the velocities with which electrons revolve in the molecule, and generally of an electronic theory of matter : but in absence of knowledge the mass of an electron was then naturally assumed comparable with that of a hydrogen atom. A great amount of suggestive material is to be found in Dr. Larmor's contributions to the Transactions of the Royal Society for 1894, 5, 7 ; some of them were summarised in the book called *Ether and Matter* published by the Cambridge University Press in 1900.

Suffice it here to say that the electron constitutes the basis of the whole treatment, and that there is supposed to be no true electric current except electrons in motion. They may move with the atoms, as in the electrolysis of liquids ; they may fly alone, as in rarefied gases ; or they may be handed on from one fixed atom to the next, as in the process of conduction in solids.

Conduction.

The possible modes of conduction or transmission of electricity are in fact three, which I may call respectively the bird-seed method, the bullet method, and the fire-bucket method.

The bird-seed method is adopted in liquids and usually in gases of ordinary density ; it is exemplified in electrolysis ; the bird carries the seed with it, and only drops it when it reaches an electrode.

The bullet method is the method in rarefied gases, as has been clearly realised by aid of the cathode rays: the space near the cathode represents the length between the breech and the muzzle of the gun, and the rest of the path is analogous to the trajectory of a bullet. When the projectile strikes an atom the shock may cause it to pass another on, and so continue the convection.

The fire-bucket method must be the method of conduction in solids, where the atoms are not susceptible of locomotion and can only pass electrons on from hand to hand; oscillating a little in one direction to receive them, and in another direction to deliver them up, and so getting thrown gradually into the state of vibration which we call heat. But it may be observed that this need for motion, in order to pass electrons on, becomes less and less according as the body is less subject to the irregular molecular disturbance we call heat. It may be the expansion and molecular separation, or it may be the irregular jostling and disturbance, that impede easy conduction; but certainly conduction improves as temperature falls, and transmission becomes quite easy at very low temperatures. The conduction of heterogeneous alloys is a less simple matter, being probably mixed up with back E.M.F. developed at innumerable junctions,—otherwise it would be instructive to examine the effect of low temperature on the conductivity of a metal which did not contract with cold.

The extra conductivity of hot electrolytes is a totally different phenomenon : it is not true conduction, but convective locomotion of ions, in their case. The same effect of temperature which lessens their viscosity increases their conductivity.

Insulators are bodies where conduction can only be accomplished with violence. Metals are bodies in which the transfer of an electron from one atom to another is easy, demanding no force as long as the process is not hurried—a process of the nature of a *diffusion*. The transmission of vibrations along a chain of connected molecules may well occur through a not dissimilar kind of connection; and hence the conduction of electricity and the conduction of heat, though really different processes, may have many points in common.

A fair approximation to the phenomena of conduction in metals has been worked out in detail by Riecke, Drude, Thomson, Lorentz, and many others; in which the electrons are supposed to remain free for periods so long that their mean energy of motion is a function of the temperature, as in gas-theory.

Most is known about electrolytic and gaseous conduction. In gaseous conduction the negative electrons, when free, fly fast; whereas the ions generally, and all the positive charges, travel more slowly by reason of their association with matter.

In liquid conduction charges of either sign are always associated with atoms, and travel only as ions, at a slow diffusion rate which was calculated by Kohlrausch, has been observed directly by myself,* Mr. Whetham and others, and is well known.

The rate of transmission in solids can only be inferred, and it would appear from the Hall effect (see Larmor, *Phil. Trans.*, Aug. 1894, p. 815), as if in one class of solids the positive were able to travel fastest, whereas in another class negative travelled fastest : a difference which is familiar in liquids. In acids, for instance, the positive charges travel much the quickest, because they are associated with light

* Lodge, *British Association Report*, Birmingham Meeting, 1886, pp. 389-413.

and presumably small hydrogen atoms; and it is
owing to the comparatively easy migration of the
light or small hydrogen atom that acids are in
general such good conductors.

The Hall magnetic bend, like Faraday's magnetic
rotation, is a differential effect, and would be zero
if positive and negative were equally acted upon. In
gases it is differential too, but there the negative
charges are liable to be so free as compared with
the positive, and to be so conspicuous, that the Hall
effect in gases, especially in rarefied gases, is very
great in comparison with the small residual effects
found in liquids and solids. Consequently the
effect of a magnet in curving the path of cathode
rays in a Crookes tube is readily demonstrated.

Radiation.

But it is not only the progressive motion or locomotion of the electric atomic appendages that we
have to consider; we must assume also that they
are susceptible of motion in the atom itself, either
vibrating like the bead of a kaleidophone, or revolving in a minute orbit like an atomic satellite.
Indeed it is to the concerted vibrations or revolutions of the system of electrons, in, or on, or round,
an atom, that its radiating power is due. Matter
alone has no perceptible connection with the ether,
a fact which is proved in my paper in the *Philosophical Transactions* for 1893 and 1897;* it is
electric charge which gives it any connection, and
even then it has no *viscous* connection—there is no
connection that depends upon odd powers of velocity,
so as to be of the nature of friction,†—it is purely

* Lodge, *Phil. Trans.*, vol. 184, pp. 727-804, and vol. 189, pp. 149-166.
† See especially *Phil. Trans.*, vol. 189, p. 164.

accelerative connection; it is only when the charge
vibrates, and during its accelerative periods, that it is
able to influence the ether at a distance by emission of
waves.* These waves consist probably of alternations
of shear, with no motion of the ether as a whole,
but only a to-and-fro quiver of its equal opposite
constituents over some excessively small amplitude :
a kind of motion which constitutes what we know as
radiation. It is not the atom pulsating as a whole
which disturbs the ether, but the pulsations or
vibrations, or the startings and stoppings and
revolutions, of its electric charge. *Acceleration of
electric charge* is the only known mode of originat-
ing ether disturbance. But normal or centripetal
acceleration, involving nothing more than change of
direction, is just as effective as actual change of
speed. If an electric charge is able to describe a
small orbit four-hundred-billion times a second, it
will emit the lowest kind of visible red light.
This number of revolutions is equal to the number
of seconds in about fourteen million years, or in
the time since some early geologic period. If
it revolves faster it will emit light of higher refrangibility; and the particular kind of radiation
emitted by the atom of any substance, when in a
fairly free state, will depend on the orbital period
of its electrons, if they could be considered as independent. But if that were so, every atom would soon
radiate itself to destruction. The condition that an
atom must fulfil in order to have a chance of survival
by retaining its energy, was given by Larmor (*Phil.
Mag.*, Dec. 1897) in the form that the vector sum of
the accelerations of all its electrons, with due regard
to their signs, should be permanently null. This

* See Chaps. I. and IX., also Appendix G.

further condition is quite consistent with those imposed by dynamics.

Every frequency of rotation will correspond to a definite line in the spectrum. But if this be its real cause, radiation must be susceptible to magnetic influence, for a revolving electric charge constitutes a circular current; and if a magnetic field be started into existence with its lines threading that circuit, it must, while it is changing in intensity, cause the speed either to increase or to decrease, and so will either raise or lower the refrangibility. If electrons are revolving in every direction, and if a magnetic field is applied to them, then during the rise of the field the pace of some will be increased and of some decreased; and this increase or decrease will not stop until the magnetic field is destroyed again.

Hence it would appear that if a source of radiation is put into a magnetic field, and its lines examined with a spectroscope, they should be affected, either by way of shift or broadening, or in some other way. It happened, however, that when Dr. Larmor theoretically perceived this, and estimated the order of magnitude of the effect to be expected, he made the assumption natural in 1895 that an electron was comparable in mass with a hydrogen atom. On this assumption, knowing what he did about the massiveness of an atom, he calculated that the effect would be too small to see; indeed, Faraday had, with imperfect appliances many years ago, looked for some such effect—not then guided by theory, but simply with the object of trying all manner of experiments—and had failed to see anything; Prof. Tait also had been moved by theory in the same direction, but no fresh experimental attempt to examine the question was initiated. Nor was the matter publicly referred to until, as hinted above, Zeeman of Amsterdam, in 1897, with a good grating and a strong electromagnet, skilfully observed a minute effect, consisting in a broadening of the lines emitted by a sodium flame placed between its poles. On seeing a two-line notice of this in *Nature* in December, 1896, Dr. Larmor wrote to me, saying that this must be the effect which he had thought of, but concluded must be too small to see. On receiving this intimation I immediately, with a little trouble, repeated and verified the experiment,* and exhibited it at the Royal Society soirée in May that same year.

From this simple but important beginning the large subject of the influence of a magnetic field on the radiation from different substances has been laboriously worked at; not only by the original discoverer, but by Preston in Dublin, Michelson in America, Runge, and others; and a whole series of important facts have been made out. Every line has been studied separately; some lines are tripled, some quadrupled, some sextupled, and so on—as said above. One mercury line is resolved into nine or perhaps eleven components. The effect is therefore *not* too small to see, though it needs excessively high power and perfect appliances to display it; and so it became evident

---

* See *Proc. Roy. Soc.*, vol. 60, pp. 466, 513, and vol. 61, p. 413, or *Nature*, vol. 56, p. 237; also several articles by Lodge in *The Electrician*, for 1897, vol. 38. The whole matter is elucidated by Zeeman, aided by Lorentz, on the basis of theory illustrated by a picture or model of an orbitally revolving electron, which, though crude, was adequate as a guide : the small mass of the revolving particle being thereby deduced, and being in general conformity with J. J. Thomson's direct determinations of the mass of an electron some months previously. With higher experimental power greater precision was reached, and an unexpected development appeared in the tripling of each line, a result which was suggested by the model, but could not have been predicted from it alone. Other lines were found to divide into more than three components, in a very suggestive but still imperfectly understood manner.

that if radiation were due to moving electrons, their motion could not be handicapped by having very much matter associated and moving with them. It became possible, indeed, by making a measurement of the *amount* of doubling undergone by the lines in a given field, to ascertain how much matter was associated with the revolving electric charge in any given case; in other words, to make a determination of the electrochemical equivalent effective in radiation—*i.e.*, of the ratio m/e. Indeed, Professor Zeeman, with considerable skill, had made a rough determination of this kind at a very early stage, when he only saw the effect as a slight broadening of the sodium lines; and had come to the conclusion that the electrochemical equivalent was quite different from that appropriate to electrolysis, being some thousand times smaller. He found, in fact, that the ratio e/m had in this case also the notable value already suspected in connection with cathode rays, viz., the value 10^7 c.g.s.

More recent measurements have confirmed this estimate, and shown that the ratio of charge to matter in the Zeeman case is practically identical with the ratio of charge to matter in the cathode ray case; in other words, that whatever is flying in the cathode rays is vibrating in a source of radiation; and that if the cathode rays consist of moving electrons, radiation is due to vibrating or revolving electrons. The more the details of the Zeeman effect are studied, the clearer it becomes that the electron theory attributed to it from the first by Zeeman and H. A. Lorentz, as well as by FitzGerald and Larmor in England, is satisfactory, though not as yet fully and completely worked out.

One of the earliest publications in England, both

of the fact and of its elementary theory, is that given by the present writer in two articles in the *Electrician* for February and March, 1897,* which are worth referring to as representing incipient ideas on the subject before the full significance was grasped. The high value of the e/m ratio, viz., $\frac{1}{2} \times 10^7$ c.g.s., or fifty million coulombs per gramme, instead of the moderate electrolytic value, is spoken of on page 643 as a difficulty; and a FitzGerald suggestion amounting virtually to the beginnings of an electron theory of the Zeeman effect is hinted at. Likewise an extremely short way of expressing the theory of the motion is given by the writer, in the following form :

Consider the resolved part of any orbital motion projected on to a plane normal to the applied magnetic field H ; and let the angular velocity be ω, at any point of an orbit where the radius of curvature is r ; then the field will exert a radial component—

$$\pm \mu e H r \omega,$$

which will represent an increment or decrement of centripetal force

$$d\left(mr\omega^2\right) ;$$

whence it follows, to a first approximation of order of magnitude, that—

$$d\omega = \pm \frac{\mu e H}{2m},$$

and the change of frequency caused by the magnetisation in the transverse components of the radiation will therefore be—

$$\pm \frac{\mu e H}{4\pi m}$$

The other or longitudinal component of the original orbit will manifestly be unchanged. This is far from

* See Lodge, *Electrician*, vol. 38, pp. 568 and 643.

being a complete and satisfactory theory, unless the projected motion happen to be circular ; but it was a brief and early attempt.

An instructive and interesting method of demonstrating the Zeeman effect was devised by W. König and described in the Jubilee volume of Wiedemann's *Annalen*, 1897, of which a brief abstract is given in *Nature*, vol. 57, p. 402. An emission flame containing the salt under examination is placed in a strong magnetic field, and viewed through an absorption flame, containing the same salt, by means of a doubly-refracting prism, or other double image instrument, so as to get two images of the emission-flame side by side. On exciting the magnet, the emission frequency, of vibrations perpendicular to the lines of force, is put out of tune with the absorption frequency, and accordingly the amount of absorption is much diminished. The result is that one of the images brightens up every time the magnet is excited ; the other image, which corresponds to vibrations along the lines of force, remaining unchanged and constituting a convenient standard of brightness.

CHAPTER XI.

FURTHER DISCUSSION OF THE ELECTRON THEORY OF THE MAGNETISATION OF LIGHT AND DE-TERMINATION OF THE m/e RATIO IN RADIATION.

AMONG the early contributions that have been made to the theory of moving charges, few are more remarkable than those of Dr. Johnstone Stoney in connection with the process of radiation, long before there had been any experimental verification of the separate existence of these electrons, or of the fact that the emission of light from a substance is due to their motion. Dr. Stoney had treated them in an astronomical manner, in 1891, dealing with an electron moving round an atom as if it were a satellite moving round a planet, and had discussed the various perturbations to which they might be subject, and the effect of those perturbations on the spectrum of the light emitted.*

One of the simplest kinds of perturbation, fully analysed by Newton for the motion of the moon, is what is called a progression or recession of the apses, being a slow revolution of the orbit in its own plane. Such a motion was shown to be able to

---

* "On the Cause of Double Lines and of Equidistant Satellites in the Spectra of Gases." G. Johnstone Stoney, *Transactions of the Royal Dublin Society*, iv., 1888-92, pp. 563-608.

originate a doublet in the spectrum; for of the two component circular vibrations into which the motion can be analysed, one has been made more rapid and therefore its light raised in refrangibility, the other has been made slower and therefore lowered in refrangibility.

Another closely allied kind of perturbation, analogous to precession of the equinoxes in the case of the earth, would result in a triplet in the spectrum. This precessional motion occurs in an orbit subject to any oblique pull or deflecting force. Instead of yielding directly to that pull, its effect is to make the axis describe a kind of cone, the kind of motion that one sees in an inclined spinning-top : the pull of gravity on a spinning-top does not make it topple over, but makes it precess. So also with a hoop or bicycle when not vertical : instead of tumbling, it turns round and round in a circuit, as long as its motion continues; only falling when the motion ceases, or falls below a certain critical value. Hence if the orbit of an electron were subjected to an oblique or deflecting force, the effect would be, not to place it directly in the desired position perpendicular to a line of force, but to cause it to precess. And this motion might be analysed into three components,—the accelerated and retarded circular orbits above-mentioned, which would result in a doubling of the line, and a third component, viz. the one parallel to the axis, which would be unchanged and would therefore represent a spectral line in its old position, the centre of a group of three. All this was clearly perceived in connection with Dr. Zeeman's discovery, with the assistance of his great compatriot the eminent physicist, H. A. Lorentz; whose theory was in several respects anticipatory of the experimental results.

It must be observed that the light emitted by the oscillation-components, above spoken of, will be all of one definite kind, due to vibrations in one definite direction, and will therefore be polarised. The kind of polarisation must depend on the aspect from which the light is seen. If seen at right angles to the axis of precession, all three lines should be plane polarised—the middle line at right angles to the other two. If, however, it be looked at along the axis of precession, then there should be no middle line; because the axial vibration would then be end-on, in which direction it produces no optical effect; and the two side lines would be circularly polarised.

Fig. 15 consists of diagrams illustrative of the changes caused in a spectrum line by application of a powerful magnetic field to the source of radiation.

I represents a specially simple case. The cadmium line A, seen by rays travelling along the lines of force, resolves itself into two lines B and C which are circularly polarised in opposite directions. This is due to the acceleration of one circular component of the rectilinear or elliptical vibration and the equal retardation of the other component.

II represents the same simple line seen by light travelling across the lines of force. In that case the line becomes triple; and if A had been plane polarised, B and C are polarised in a plane at right angles to that of A'. This is due to a precessional movement of the plane of the orbital motion, the axial vibration continuing unchanged, and the two at right angles being one accelerated and the other retarded.

III and *IIII* represent the effect of a magnetic field, applied to a sodium source, on the constituents of the yellow sodium double-line. D_1 is resolved into a quartet, and D_2 into a sextet, when the light travels across the magnetic field.

Directly Zeeman had demonstrated the fact that a magnetic field applied to a source of light was able to act as a perceptible perturbing cause, Professor Lorentz was at once able to predict the main part of that which has been here stated,—about the tripling of the line seen sideways to the lines of force, and the doubling of the line seen endways,—with all the polarisations as just stated ; because the lines of

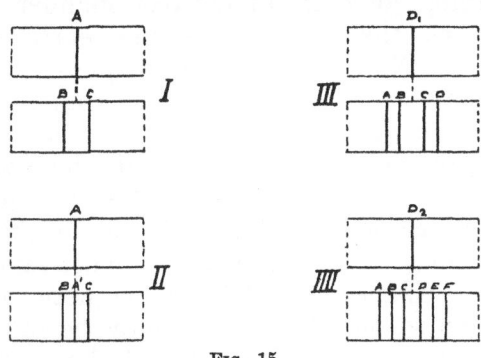

FIG. 15.

magnetic force constitute the precessional axis. And all these effects were shortly afterwards seen by Zeeman and others, and are characteristic of the simplest circular orbit.

As already stated, the full meaning of these very exquisite phenomena is still very far from being unravelled. The most general theoretical result is that of Larmor (*Phil. Mag.*, Dec. 1897) that for any atomic system, however complex, if the effectively moving electrons are all negative, while the attraction of the positive on them is centrical, each line will be divided into three, exactly as in the provisional theory of Zeeman and Lorentz.

At first sight one might be inclined to suppose that the orbits would all face round and set themselves normal to the lines of force, like so many circular currents; but that is to forget the inertia of the travelling electron. It is manifest that since a revolving electron constitutes a circular current, its *tendency* will be to set itself with its plane normal to the lines of force; but since by hypothesis the revolving electron has inertia, the current will not so set itself, but will yield to the deflecting force in an indirect manner, as a top does; or as the oblate spinning earth does—as explained by Newton in the Principia,—the axis of rotation having a conical motion round the lines of force : a motion which is called "the precession of the equinoxes" in the case of the earth, and "the Zeeman effect" in the case of a radiating atom.

This is an account of the chief part of the Zeeman effect, and may be regarded as the most fundamental kind of disturbance caused by a magnetic field on a source of radiation. But there may be other minor disturbances, just as in the case of the earth, whose axis is not only subject to precession, but also to nutation—a nodding movement superposed upon the main motion. It is also quite possible for the middle line, or for the two outer lines, or indeed for all three lines, to be doubled; thus giving rise not to the standard triplet, but to a quartet or a quintet or even a nonet,—appearances seen and photographed by Zeeman, Preston and others. The remarkable 'echelon' spectroscope of Michelson has been invented just in time for application to phenomena of this kind—its special function being the close examination in detail of a minute portion of a spectrum otherwise produced.

Even the two constituents of the double sodium line behave differently, when the source is magnetised and the light thus examined—as illustrated in fig. 15 : one of the sodium lines, D_2, which had appeared only broadened to Zeeman at first, really becomes a sextet. The other sodium or D_1 line becomes a quartet ; and a complete study of the behaviour, under magnetism, of all the lines and groups of lines given by different substances must result in a great extension of our knowledge in many directions ; in fact it is hardly too much to say that the discovery of Zeeman, in the light of the theory of Lorentz, has doubled the power of spectrum analysis to throw light upon the processes of radiation and the properties of atoms, and has opened up a new branch of physics—a new department, as it were, of atomic astronomy, with atoms and electrons instead of planets and satellites.

CHAPTER XII.

INCREASE OF INERTIA DUE TO VERY RAPID MOTION.

THE hypothesis to which we have been led is that the inertia of an electron is wholly of an electrical character, and is explained by the known magnetic effect of an electric charge in motion, and the consequent reaction to any change in that motion.

Usually inertia is treated as constant and quite independent of speed; but now arises the question whether the distribution of charge on a charged body, together with its lines of force, will remain constant and unaltered while the body is rapidly moving; because if the distribution of lines of force is altered, then the inertia due to their lateral motion will probably be altered too. This can be made plain after referring back to Chap. II.

Thus, for instance, imagine that the lines of electric force of a body in motion became more concentrated towards the axis or line of motion; the effect would be at once to diminish the lateral component of their motion, therefore to diminish the magnetic force which that lateral component causes, and thus to diminish the apparent or electromagnetic inertia of the moving charge.

On the other hand, if the lines opened out and became concentrated towards the equator, or plane normal to the line of movement, then a greater component of their motion would be of a kind suitable to excite a magnetic field; moreover, both the fields would by this concentration increase in intensity, and the inertia would increase.

Thus, then, it may be possible that electric inertia may depend in some fashion on speed, a thing unknown in ordinary mechanics. I do not say that such dependence must be *untrue* in ordinary mechanics; on the contrary, I feel reasonably sanguine that it will be found true for matter also, when moving sufficiently fast—say over a thousand miles a second,—though it is unlikely that it can have a practical influence in any actual known case of rapid movement in astronomy. But however this may be, there is no doubt that theory points to an increase of electro-magnetic inertia at excessively high speeds, and Mr. Heaviside long ago calculated its amount.

It will be observed that when a charge moves, it generates circular magnetic lines of force. Now these magnetic lines are not stationary, but are themselves moving at the same rate as the body; hence they generate fresh electrostatic lines, *i.e.*, cause an electric displacement away from the axis, which displacement is superposed upon the original radial displacement (away from or toward the centre) due to the charge.

At ordinary, at even violent speeds, this second-order electric effect is insignificant, but it is there all the time, and must not be ignored when the speed becomes extravagantly high. It rapidly rises into prominence when the speed approaches the velocity of light, but at any speed much smaller than this such a second-order effect is negligibly small.

Its effect will be, as the annexed figure shows, to alter the arrangement of the lines of force, making them move away from the poles and concentrate towards the equator of the charged sphere, when the speed is very great; ultimately becoming wholly concentrated upon, or parallel to, the equatorial plane, in the limit; if the speed could attain that of light. And the electric lines of force would then be opened out into a fan or equatorial brush, like the spokes of a wheel which is rushing furiously along an elongated axle, the circumference of the

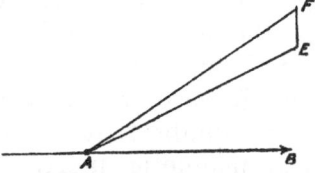

FIG. 16.—A is the charge, AB its line of motion, and AE its electric force in a certain direction when stationary ; EF is the magnetically induced electric component due to the motion and AF is the resultant electric force which replaces the original force AE. The magnetic force, to the motion of which EF is due, is perpendicular to the paper, and is itself caused by the motion ; hence EF is a quantity of the second order and is small for speeds distinctly less than that of light.

wheel representing the direction of the magnetic field; but this very condensation so intensifies the field as to make the inertia ultimately infinite.

It might be supposed that rearrangement of the lines means that the distribution of the charge itself is altered by the motion, so that all the charge is concentrated upon the equator, whence the lines of force would start normal to the surface as usual. There are many difficulties about such a conception however (see Appendix K), and it is easier to suppose that the charge retains its distribution un-altered, on the surface of the sphere, and that each line of force starts from its original point; but that it starts no longer in a direction perpendicular to the surface, when it is in rapid motion, but sets out obliquely, with a deflexion towards the equator, so as to give the arrangement above described ;—like trunks of trees on a cliff or landslide which preserve their roots *in situ* and gradually adjust their growth to the vertical direction without being any longer perpendicular to the soil.

As a matter of fact the question all depends on what hypothesis we make as to the intrinsic structure of the electron. If we liken it to a perfectly conducting body with an electric charge, the charge must be confined to its surface; and it may be proved, as Heaviside did, that the distribution will remain uniform (cf. Larmor, *Æther and Matter*, p. 154). Or it might be likened to a solid globe of uniform electrification. It may be something of which we have as yet no conception : but the experiments of Kaufmann probably suffice to prove that, whatever the structure is, it is symmetrical around a centre, after the general manner of a stratified spherical distribution.

On the other hand these considerations can be avoided by treating the charge merely as a geometrical point from which the lines of force emanate, and ignoring its size or possible conducting power. This is the keynote of Larmor's treatment throughout his book *Æther and Matter*, and also in his earlier papers : in dealing with atomic structure it implies that the electrons in the atom are at distances apart which are great compared with their radii. Cf. the fundamental investigation of Chapter XI. to be referred to below. We could hardly tell *a priori* which treatment would best correspond with fact, but it will turn out (see Chap. XIII.) that this second method of treatment is not only simpler but that it is adequate to existing knowledge, enabling numerical results to be obtained which are singularly concordant with experimentally measured results.

In any case an indication of the mode of attack can be suggested thus :

The magnetic force due to motion is proportional to the speed of the motion. The secondary electrostatic force due to the motion of this magnetic field is likewise proportional to the same speed.* Hence the disturbance of the original uniform electrostatic field will be of the second order, u^2/v^2; and when-

---

* The value of the magnetic force at any point P, with polar coordinates $r\theta$, due to a charge e flying with speed u, is

$$\mathbf{H} = \frac{eu \sin \theta}{r^2},$$

and is in rings round the line of motion u. It is not shown in the diagram because it is perpendicular to the paper, through P.

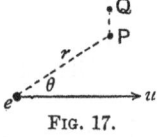

Fig. 17.

The electric force generated or induced by motion across this magnetic field—which is necessarily at right angles to the direction of motion—is $\mu \mathbf{H} u$; and in this case is therefore equal to

$$\frac{\mu e u^2 \sin \theta}{r^2}.$$

This is the secondary or induced E.M.F., to be superposed at every point on the primary or direct electric force of the charge itself along the line eP, namely $e/\kappa r^2$; and it is in the direction PQ, being perpendicular both to the magnetic field and to the motion. So the ratio of the induced to the original E.M.F., at every point in a direction θ, reckoned from the charge and axis of motion, is

$$\mu \kappa u^2 \sin \theta, \quad \text{which equals } \frac{u^2}{V^2} \sin \theta.$$

In consequence of this the original direction of the stationary electric field, eP, is displaced or tilted into a position such as eQ.

ever we can afford to neglect quantities of this order, the field and therefore the inertia of the moving charge will continue practically constant.

But when its speed of motion begins to approach the velocity of light, say even no more than $\frac{1}{10}$th of that speed, then a perceptible disturbance is to be expected, and something like a 1 per cent. increase of inertia must occur.

The complete investigation makes the inertia infinite when the speed reaches that of light (see Appendix K), but there is probably no need to press this to extremes, unless the charge were an absolute point; clearly, however, the inertia will then be very great, and possibly therefore it may always be impossible to make matter, or at least charged matter, move with a speed greater than that of light. There may be ways out of this, however, just as it is possible for a bullet to move through air with a velocity greater than that of sound. This is managed by the violent adiabatic condensation of the air in front of such a bullet, the effect being to raise the appropriate velocity of sound to the required value; and by the ridge behind it where discontinuity makes its appearance. It seems unlikely that the ether can adjust itself to excessive speed beyond the speed of light without a change of structure akin to what would be rupture in the case of a material medium.

It has been shown both by Mr. Heaviside and by Prof. J. J. Thomson that if the speed of motion *is* ever greater than that of light, the fan or radial plane of lines of force bends backwards and becomes a conical surface, gradually closing up as the speed further increases: in accordance with the analogy of the conical surface of discontinuity aforesaid, which travels with a sufficiently rapid bullet, and is demonstrated in Mr. Boys' bullet photographs.

No known speed which exists in ordinary matter is sufficient to bring any variation of inertia into prominence. The quickest available carriage is the earth in its journey round the sun, 19 miles a second, or 60 times faster than a cannon ball; but the earth's velocity is only the $\frac{1}{10000}$ of the speed of light, and consequently any spurious inertia due to its orbital motion is only 1 part in a hundred million; and even the accuracy of astronomy could not display an effect of that order of magnitude.

There are a very few stars which move 200 miles a second, but even these have only one-tenth per cent. of the speed of light, and the excess inertia will be only 1 part in a million. The only known place where charges or charged atoms were known, prior to 1903, to move at speeds greater than this, was in a vacuum tube. There the cathode-propelled particles are flying 20,000 miles a second or $\frac{1}{10}$th the speed of light, and they may have 1 per cent. excess inertia; or more if they can be persuaded to go still faster.

But higher speeds are now known, being obtained in the spontaneous emission of electrons and atoms by radio-active materials; so it becomes of the greatest interest to determine the constants, and especially the inertia, for rays of this kind.

CHAPTER XIII.

JUSTIFICATION FOR ELECTRIC THEORY OF INERTIA.

But first we must ask, what justification is there for the view that each of the isolated corpuscles, on which measurements have been made, is a purely electrical corpuscle or electron without material nucleus, all of whose properties are to be explained in accordance with purely electric and magnetic laws? Then we may proceed to discuss the further extraordinarily far-reaching hypothesis—first tentatively put forward by Larmor in 1894, *Phil. Trans.*, vol. 185A, p. 813, with mechanical illustration of a purely ethereal structure for such an electron—that the electrons constitute matter, that atoms of matter are composed of electric charges, that the fundamental inertia-property of matter is identical with self-induction.

There is the reasonable philosophical objection to postulating two methods of explaining one thing. If inertia *can* be explained electrically, from the phenomena of charges in motion, it seems needless to require another distinct cause for it also. But this is not all that can be said; it is quite possible that direct experimental proof will be forthcoming before long. One method suggested by Professor J. J. Thomson, for examining the nature of the corpuscles, had reference to the proportion of radiation to thermal energy developed when corpuscles encounter a target which suddenly stops them.   In so far as they consist of non-electric matter they would produce only heat by their dead collision, without any direct generation of ethereal waves ; in so far as they consist of electric charges they would disperse a certain amount of radiation energy ; and so the proportion of radiation to heat might afford a criterion.*   Hitherto, however, no adequate measurements have been made in this direction.

But there is another more likely avenue to a conclusive result.   We have seen that when an electric charge moves with a speed approaching that of light, its inertia is theoretically no longer constant, but rapidly increases and becomes infinite when the light-velocity itself is reached; and rather complicated and different expressions for this high-speed inertia have been calculated by several mathematical physicists, on different views of the constitution of the electron.   See Appendix K for a discussion of this difficult subject.   It is possible that this fact will give us the necessary clue.

For in certain cases of the production of cathode rays, or at any rate of beta rays, a speed not far short of that of light is reached, and in such cases the effects of the increased inertia can be observed. Such an experimental determination has been quite recently undertaken and executed with great skill by Dr. Kaufmann,† who employed the method indicated above (Chap. V.) of comparing simultaneously the electric and the magnetic deflexion of the same set of rays from a speck of radium submitted

---

* See J. J. Thomson, *Phil. Mag.*, April, 1899, p. 416.
† See *Comptes Rendus* for October 13, 1902.

CH. XIII.]   VARIATION OF INERTIA   131

simultaneously to an electric and a magnetic field coincident in direction. As a matter of fact the speck gives off rays of various speeds, which are differently deflected into a thin streak like a comet's tail (see fig. 18): and it is the faint impression they make on a photographic plate in high vacuum that is measured and gives the data. Thus the velocity and the e/m ratio are both known, and—to summarise briefly the result— Kaufmann concluded that when the speeds approached perceptibly near the velocity of light, the electrochemical equivalent m/e increased by just the amount required in accordance with pure electric theory—the theory which attributes the whole of inertia to electric influence. There appeared to be no quantitative room for any extra inertia, such as that of an inert particle of non-electric matter travelling with each projectile, retaining its inertia constant at all speeds, and so contributing nothing to the rise of inertia perceived when the speed approaches within hail of that of light.

We will now enter more into detail concerning this important matter.

Proof of the purely electrical nature of the inertia of the β particles shot out by Radium.

There is every reason to believe that the β rays emitted by radium are identical with the cathode rays observable in a vacuum tube ; for both consist of a multitude of electrons or corpuscles travelling at excessively high speed ; and if a determination be made of this speed and of the electro-chemical equivalent for the case of β rays—for instance, by the method of subjecting them both to magnetic and to electrostatic deflexions, which is the easiest way—the numbers come out quite similar to the number obtained for cathode rays, viz., for m/e the value 10^{-7} in E.M. units, and for u something of the order 10^9 centimetres per second.

But radium under favourable conditions is found to shoot out its particles with a speed exceeding even this, and in some cases to approach within hail of the limiting speed, the velocity of light. This is the very important result obtained by the German physicist W. Kaufmann, who has made an admirable series of determinations of speed and of electro-chemical equivalent for this case. The importance of obtaining these excessively high speeds should be obvious, for thereby we are enabled to test the electrical theory of inertia. Theoretically the inertia at high speeds is not constant, but increases according to a complicated but calculated law; we cannot suppose that the electric charge varies in any way with motion; hence the electrochemical equivalent m/e is proportional simply to the mass, and ought to be a function of the velocity u, nearly constant for ordinary values, but increasing rapidly as it approaches within hail of the velocity of light.

To obtain numerical values we may apply the theory developed by Mr. Heaviside and by Prof. J. J. Thomson, with regard to the increase in momentum of a flying electric charge, over and above the natural mu value, with m considered constant, which is the value at all ordinary speeds.

The formula which the latter used for the purpose of numerical calculation is one of those given in his *Recent Researches*; it is quoted in his *American*

Lectures on Electricity and Matter, p. 44, as follows :

$$mu = electric\ momentum =$$

$$= \frac{e^2}{2a} \frac{v^2}{(v^2 - u^2)^{\frac{1}{2}}} \left\{ \theta \left(1 - \frac{1}{4} \frac{v^2}{u^2} \right) + \frac{1}{2} \sin 2\theta \left(1 + \frac{1}{4} \frac{v^2}{u^2} \cos 2\theta \right) \right\}$$

where $\sin \theta = u/v$; and we may express it in a fairly simple form thus :

The momentum of a particle of electricity moving at excessive speed is greater than the momentum of the same particle estimated on the hypothesis that its mass is constant, in a numerical ratio given by the following expression ; where the ratio of the speed to that of light, u/v, is expressed as the sine of a certain angle θ :

$$\frac{\cdot 75}{4 \sin^2 \theta \sin 2\theta} \left\{ (4 \sin^2 \theta - 1) 2\theta + (4 \sin^2 \theta + \cos 2\theta) \sin 2\theta \right\}.$$

We will call this the ratio $\phi(\theta)$. It is the measure of the spurious or extra inertia due to rapid motion ; the ratio of the mass at speed u to the stationary mass. We may also write it, rather more conveniently perhaps for calculation, thus :

$$\frac{m}{m_0} = \phi(\theta) = \frac{3}{8} \left(\frac{1 - 2 \cos 2\theta}{1 - \cos 2\theta} \cdot \frac{2\theta}{\sin 2\theta} + \frac{2 - \cos 2\theta}{1 - \cos 2\theta} \right). \quad \dots (1)$$

Now the highest speeds measured by Kaufmann were such as the following :

2·36, 2·48, 2·59, 2·72, 2·85 times 10^{10} cm. per sec.

while the speed of light is well known to be $3\cdot 0 \times 10^{10}$ cm. per sec. ; so the ratios u/v, corresponding to the above observed speeds, are respectively

·787, ·817, ·863, ·907, ·95.

These numbers therefore represent the values of $\sin \theta$, to be inserted in the above formula for obtaining the theoretical ratio $\phi(\theta)$; namely, the ratio which expresses the number of times the mass of an electric charge, at specified high speed, exceeds its mass at low or zero speed.

The successive values of $\phi(\theta)$ come out, according to J. J. Thomson, for the above set of velocities,

1·5, 1·66, 2·0, 2·42, 3·1,

and these are what must be compared with direct observation or measurement of the apparent or effective mass in each case.

Now the corresponding values observed experimentally by Kaufmann for these same quantities—that is to say the factor by which the moving mass exceeded the same mass when stationary—were

1·65, 1·83, 2·04, 2·43, 3·09,

showing a very remarkable degree of approximate agreement between experiment and theory,—especially at the higher speeds.

Thus at the highest speed ever yet observed for what may be called a particle of matter, at any rate for an electron—namely $2·85 \times 10^{10}$ cm. per sec. or six hundred million miles per hour—the mass of the particle is three times as great as its usual value; and naturally its momentum and energy are increased in the same proportion.

Such a surprising agreement as the above, between theory and observation, removes from my mind all reasonable doubt as to the truth of the hypothesis that the inertia of electrons is electrical inertia. I regard this closeness of agreement as specially surprising, for it was not the first deduction of the experimenter, W. Kaufmann, himself: his deduction rather was that the electrical mass constitutes about one-third or one-fourth of the whole; but then he used another formula for calculating it (given in Appendix K), which assumes that the charged body behaves like a conducting sphere. But when the correct deductions from the Heaviside expressions above referred to were applied, with the collaboration of M. Abraham, results practically equivalent to the above were obtained. The above agreement is attained by Professor J. J. Thomson, who applied his own theory to the results of Kaufmann, working it out on the assumption that the charge behaves like an actual point.

If it is to be urged in future that an electron contains a material nucleus in addition to its electric charge, the burden of proof rests with those who maintain that thesis. The hypothesis which now holds the field is the purely electrical one.

But it must be remembered that this is not the same thing as establishing an electrical theory for all matter. The inertia of an electron is purely electrical, but what about the inertia of an atom? Who knows that the atom is wholly composed of electrons? We do not know that as yet.

Nevertheless we are now in a very central chapter of modern physics, and it is desirable to enter into the matter somewhat more in detail than in the above preliminary sketch.

CHAPTER XIV.

MORE ADVANCED DEVELOPMENT OF THE COM-
BINED ELECTRIC AND MAGNETIC DEFLEXION
METHOD FOR MEASURING VELOCITY AND MASS
OF THE PARTICLES IN COMPOUND RAYS.

THE methods given in Chaps. V. and VI. and Chap.
IX., for measuring u and e/m, made the determination look very simple, if the precaution is taken of
having the apparatus *in vacuo*, so as to eliminate
the troublesome conducting power of the air and
obtain the electric deflexion undiluted, as J. J.
Thomson first found feasible. But then the simple
theory, there given, assumed that the quantities to
be measured were constant, and that the deflexion
to be observed in each case was a single deflexion
capable of accurate measurement; but this is often
far from being the case, since the velocities of the
particles differ; and when, as in the case of radium,
some of the speeds approximate to that of light, it is
impossible that it can be the case, for the inertia itself
then changes in a complicated way with the speed
and must be treated as variable. It is easy to forget
that, because it is an unusual feature in mechanics.
So the deflexion cannot be a simple deflexion, the rays
must be fanned out as it were into a spectrum (see
fig. 18); and, since this spectrum is continuous,

it will possess no features which enable anything like measurement to be made on it, unless some still further ingenious device be employed, such as, for instance, that of Kundt for making experiments on anomalous dispersion.

The experiments of W. Kaufmann at Göttingen were conducted after this very fashion, and may be summarised thus :—an electric and a magnetic field were simultaneously applied, in such a way as not to neutralise each other's effect but to cause deflexions

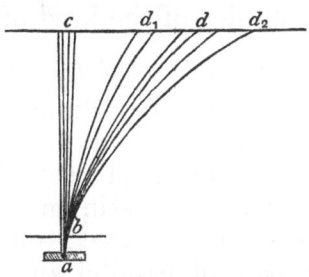

FIG. 18.—Diagram of the deflexion of high-velocity rays from radium. The radium is in a cavity in a lead block a; the rays pass through an aperture b, and are spread out by a magnetic field into a spectrum $d_1 d_2$; the gamma rays or any uncharged rays produce an impression at c on the photographic plate, cd, placed to receive all the rays. In a uniform field each of the lines abd is a circle.

at right angles to each other. In that case if the rays from a small point source, after traversing the double field, are received upon a photographic plate at a little distance, it may be expected that the two spectra will be compounded into a single spectrum inclined at some angle corresponding to the relative strength of the two fields. But whether the inclined spectrum thus produced will be a straight line or a curve must depend upon circumstances. All that can be said, without further consideration, is that each point of the spectrum would represent a definite ratio of deflexion, and therefore a definite inertia and velocity, for each of the particles which have produced the impression at that point. And inasmuch as the particles of different speeds will be sorted out to different parts of the spectrum, it may be possible to select those points which correspond to the highest speeds, and indeed to compare the ratio of the two deflexions for various speeds, if by any means the velocity corresponding to each point can be determined.

A little calculation is needed to bring out the details of the theory, and that shall be given directly, but first I will give an idea of the kind of apparatus used.

Experimental Device used by W. Kaufmann.

A minute quantity of radium salt in a little brass box acts as source, and a pencil of its rays penetrates a small hole, about half a millimetre diameter, in a plate of platinum at a distance of 2 centimetres from the source ; on the way, they pass between a pair of parallel and insulated plates of brass which are separated by about 2 millimetres from each other and connected to a high-tension battery of from 2,000 to 5,000 volts. After then travelling another 2 centimetres, they encounter the photographic plate placed to receive them. The apparatus is contained in a thoroughly exhausted vessel, and the whole is placed between the poles of a large electromagnet giving a nearly uniform field, so

Fig. 19.—Kaufmann's apparatus for measuring simultaneously the electric and magnetic deflexions of particles possessing very high velocity. The source of radiation is a minute quantity of radium placed in a box at C. All except the highest-velocity rays are deflected out of action by the magnet NS; some of the highest-velocity rays pass upwards through the aperture D, being deflected forward by the magnetic field, and sideways by an electric field, whose lines are coincident with the magnetic lines, between the adjustable plates P_1P_2, which are kept as highly electrified as possible through the electrodes R. These thus doubly-deflected rays then fall upon the photographic plate E, where they are spread out into an oblique sort of very minute spectrum, more or less in accordance with diagram 18; on which spectrum micrometric measurements are subsequently made.

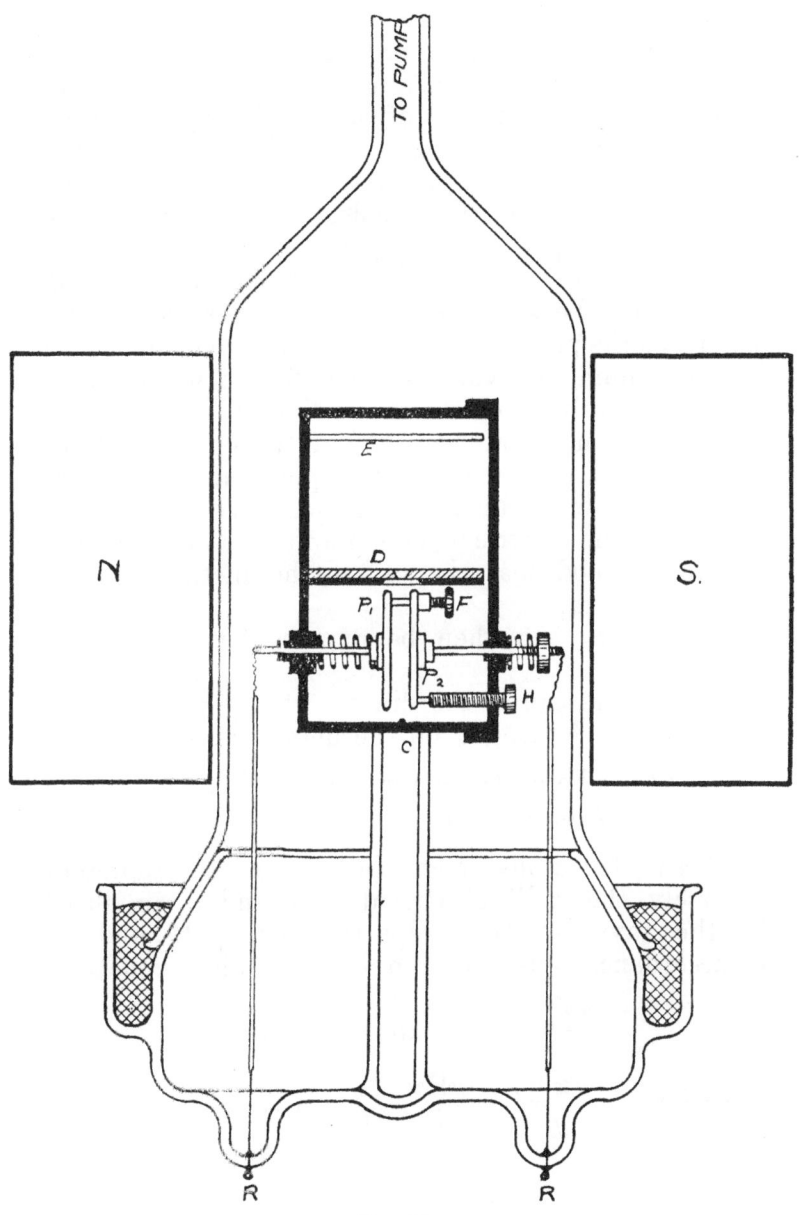

FIG. 19.

that the magnetic and electric fields are superposed in the same direction, their lines of force being coincident.

Under these circumstances the particles will describe the beginning of a spiral, being curved round the magnetic lines and deflected along the electric lines, until they escape from the combined field and travel in their deflected direction to the photographic plate as target. The slow-moving particles, if any, will presumably strike the bounding surfaces and be stopped,—only the very rapid ones will reach the plate; which is protected from alpha-rays by aluminium foil, while the undeflected gamma-rays would probably mark the direct line of fire, and thus give the geometrical "origin" of the curve or trace which would be found on the plate after long exposure,—a curve which we may write $y = f(x)$, where y signifies the electric deflexion and x the magnetic.

This method may be called the method of the crossed spectra.

The theory can then be expressed somewhat as follows :

Let the measured coordinates of any point in the spectrum, as developed on the photographic plate, be x and y ; x being the magnetic deflexion, and y the electric.

These deflexions may be taken to represent inversely the radii of curvature r and r' produced in the rays by the respective fields H and E, in accordance with the simple mechanical equations

$$\frac{mu^2}{r} = \mu e \mathrm{H}u, \text{ and } \frac{mu^2}{r'} = \mathrm{E}e,$$

wherefore $\dfrac{x}{y} = \dfrac{r'}{r} = \dfrac{\mu \mathrm{H}u}{\mathrm{E}} = \dfrac{u}{k_1}$, ...(1)

where k_1 is a constant depending on the relative effective strengths of the fields applied.

In so far, therefore, as the particles which reach the plate are all emitted with nearly the same velocity, the photographic trace will be an approximate straight line, whose slope is a measure of that velocity.

But to get the electrochemical equivalent we must also write, from the above equations,

$$\frac{m}{e} = \frac{\mathrm{E}r'}{u^2} = \frac{\mu^2 \mathrm{H}^2}{\mathrm{E}} \cdot \frac{r^2}{r'} = k\frac{y}{x^2}, \quad \ldots\ldots\ldots\ldots\ldots(2)$$

where k is another constant expressive of experimental conditions; so in so far as the masses of the particles are all the same, the photographic trace or spectrum will be a parabola.

But at the highest speeds m/e is not a constant, but a function of u,—such a function as is given on page 133,—with $u = v \sin \theta$.

So calling this function $\dfrac{m}{m_0} = \phi\left(\dfrac{u}{v}\right) = \phi\left(\dfrac{k_1}{v}\dfrac{x}{y}\right)$ we arrive at the conclusion that the actual equation to the photographic curve should be

$$\frac{x^2}{y}\phi\left(\frac{k_1}{v}\frac{x}{y}\right) = k_2,$$

with k_2 another constant.

At the highest speeds, when u approaches v the velocity of light, u cannot vary much, since it is approaching a limit, and accordingly the curve to be expected will be approximately a straight line; the only rapid variable will then be the mass, which is getting near to its asymptotic approach to infinity, and therefore varies much more rapidly than u.

The determination consists therefore in getting as

clear a trace as possible, for purposes of measurement, and then by trial and error choosing a constant k_1, such as to make the squares of discrepancies of the ratio here called k_2, from its mean value, as small as possible.

If it is possible to find a value for the constant k_1 which shall bring out the calculated value k_2 constant within the limits of experimental uncertainty, then the form of the theoretical function ϕ is to that extent verified; and inasmuch as that function was calculated on the hypothesis of purely electrical mass, the hypothesis of the purely electrical nature of the inertia of β rays is thereby similarly verified.

Kaufmann in one of his papers says the experimental errors in his concluding series only amounted to 1·4 per cent.; which, considering the difficulties to be overcome, is remarkably good.

It is also of interest to record that the numerical value obtained for the normal or low speed value of $\dfrac{e}{m_0}$ for the β rays from radium is $1·84 \times 10^7$ c.g.s.; while Dr. Simon's independent determination, by other means, of the same quantity for cathode rays was $1·865 \times 10^7$; which is likewise a satisfactory agreement. It is needless to emphasise the agreement with J. J. Thomson's much earlier measurement of the same quantity for rays from other sources.

The formula employed by Dr. Kaufmann, as representing the inertia, was erroneously deduced from results in a paper by Mr. Searle of Cambridge; and on the strength of that he concluded at first that only a fraction of the mass was electric. But it was pointed out by Dr. Abraham of Göttingen that the inertia thus calculated was only appropriate to direct acceleration, or acceleration in the line of motion; whereas what was wanted was the transverse inertia, or the inertia appropriate to a deflecting force at right angles to the line of motion.   This is to be obtained from the expression for the tranverse force, derivable from the expression for the energy in the ordinary manner by Lagrange's dynamical equations : at high speeds its value comes out different; and when the formula supplied for it by Dr. Abraham was subsequently applied by Kaufmann in his calculations, it was found to correspond very nearly with the view that the whole of the inertia is electric.

This formula, which in fact applies to any solid aggregation of electricity stratified spherically, is that the transverse inertia of a flying particle, m, is to the inertia of the same particle stationary or moving slowly, m_0, in the following ratio :

$$\frac{m}{m_0} = \frac{3}{4}\psi(\beta) = \frac{3}{4\beta^2}\left(\frac{1+\beta^2}{2\beta}\log\frac{1+\beta}{1-\beta} - 1\right), \quad \ldots\ldots\ldots(2)$$

where β is the ratio of the velocity of the particle to the velocity of light.

This formula is not identical with that employed by Thomson, possibly because the latter worked with a different idea of an electron, though it gives numerical results not exceedingly different.   Primarily, however, it was employed not so much as an absolute expression, as a *form* of function to be verified : though it was used absolutely too.   Kaufmann was ultimately satisfied by finding out that his observed mass varied if anything more rapidly, not less rapidly, than theory required; so that if the particles contained any outstanding inertia of a non-electrical character, such unexplained inertia must have a negative value,—which presumably would be absurd.

I do not myself find that Abraham's function

agrees with observation, in absolute numerical value, any better than Thomson's formula does; nor do I obtain even from Thomson's formula exactly the numbers that he quotes in his American lectures; but to go into the whole matter would be inappropriate here,—nor is it necessary, since the theoretical differences only concern details that could doubtless be removed by a little discussion: it will only become necessary to go into them more fully when the difficulties of the experimental observation are still further overcome, and when even more accurate and trustworthy results are obtained.[*]

I have taken the table of Kaufmann's best results, as published in the *Physikalische Zeitschrift* 4, 1902-3, p. 55, and calculated them out by aid of the ϕ expression given above on p. 133.

The results are tabulated below. I quote his given experimental values for x and y, together with the values he gives for β, or u/v, or what I have called $\sin\theta$; and then after reckoning out $\phi(\theta)$, which represents the theoretical ratio m/m_0 according to Thomson's theory, I have put a column of y/x^2; which should correspond, at least proportionally, to the same quantity as *experimentally* determined; and I likewise quote a column of $\frac{3}{4}\psi(\beta)$, which represents the same quantity calculated according to Abraham's formula (p. 143). (The numerical agreement of y/x^2 with a mean mass ratio, without any constant factor other than unity, must be accidental.)

[*] The results of Kaufmann's *subsequent* work will be discussed in Appendix M.

Dependence of Inertia on Speed.

Analysis of Kaufmann's latest Experimental Results.

OBSERVED.				CALCULATED.					
Magnetic deflexion in centimetres. x	Electric deflexion in centimetres. y	Velocity compared with light. $\cdot389\,\frac{x}{y}$	Ratio v/v as reckoned by Kaufmann. (β or $\sin\theta$.)	Auxiliary Quantities.			Estimated ratio of the masses m/m_0.		
				arc 2θ.	$\sin 2\theta$.	$\cos 2\theta$.	J. J. T.'s Theory $\phi(\theta)$.	$\frac{y}{x^2}$	Abraham's Theory $\frac{3}{4}\psi(\beta)$.
·150	·0607	·963	·963	2·60	·515	− ·857	3·30	2·70	2·41
·175	·0720	·945	·949	2·50	·597	− ·802	2·85	2·35	2·14
·200	·0835	·931	·933	2·40	·674	− ·738	2·50	2·09	2·05
·225	·0991	·883	·883	2·165	·829	− ·559	2·00	1·96	1·73
·250	·1132	·868	·860	2·07	·877	− ·480	1·80	1·80	1·64
·275	·1290	·829	·830	1·96	·925	− ·381	1·66	1·70	1·54
·300	·1445	·807	·801	1·855	·960	− ·280	1·54	1·61	1·47
·325	·1630	·775	·777	1·78	·978	− ·208	1·47	1·54	1·41
·350	·1813	·750	·752	1·70	·991	− ·130	1·42	1·48	1·37
·375	·1988	·733	·732	1·64	·9975	+ ·070	1·35	1·41	1·35

CHAPTER XV.

ELECTRIC VIEW OF MATTER.

WHAT has been proved, by the combination of experiment and theory now summarised, is that the most important variety of rapidly flying corpuscle—the cathode rays in a Crookes tube, the β rays emitted by radium and other radio-active substances, the particles thrown off by most clean surfaces when negatively charged and stimulated by ultra-violet light, the carriers of the negative discharge from incandescent bodies, and likewise the revolving or vibrating portion of an atom to which the emission of radiation or ethereal wave-motion is due—that these are all of an identical nature and all possess an inertia of purely electromagnetic character ; that is to say that they are all pure electrons without any admixture of ordinary or unexplained matter,—that they are simply electrical charges without any material or non-electrical nucleus.

We thus reach the very important deduction that negative electricity can exist apart from matter in little isolated identical portions, each of exceedingly minute known size, known charge, and known inertia; and that the laws of mechanics, applied to such particles in given fields of electric and magnetic force, should carry us on towards explaining the fundamental phenomena of electric currents, of magnetism, and of the production of light. But in order to explain chemical action, the details of radiation—such as groups of spectrum lines and the like,—the differences observable between conductors, and the properties of magnetic bodies, it is necessary to treat of *matter* also ; and to consider whether its inertia too, and therefore its whole nature and properties, can be reduced and simplified and explained as electromagnetic phenomena.

For observe that though an electron has been shown to possess purely electrical inertia, the same proof has not yet been extended to an atom : the constitution of an atom is so far unknown, and is the subject of hypothesis only. Moreover the only electron observed, so far, has been the negative electron ; the positive has hitherto escaped observation in any isolated form, since it has never been met with apart from a mass comparable with an atom in bulk and weight. It may be that it can have no separate existence apart from the atom of matter, but in that case it will hardly be proper to speak of it as an electron at all ; it may be that an indivisible positive charge itself constitutes the bulk of an atom of matter ; in any case its nature must be investigated, and many have been the attempts made in that direction,—among the best known in recent times being the experiments of Prof. Wien and others on "Canal rays." According to Larmor positive electricity *must* be the mirror image of negative, and experimental results must be interpreted to suit that theoretical conclusion. The relations of positive electricity constitute in fact the main outstanding problem of Physics at the present time, and until they can be probed, further progress

towards understanding the constitution of an atom
must remain in a state of suspended animation.

The only portion of an atom that has been really
analysed, and so to speak understood, is that minute
but significant fraction of its mass which confers
upon it an electric charge, consequent chemical
affinity, and radiating power ; and, when we come to
ask what all the rest of the atom is composed of, all
we can say definitely is that the specific structure
must depend on the nature of the chemical element
under consideration.   But if we consider the simplest
known atom, namely that of Hydrogen, we can make
various hypotheses somewhat as follows :

(1) The main bulk of the atom may consist of
ordinary matter (whatever unknown entity is hidden
by that familiar phrase), associated with sufficient
positive electricity (whatever that may be) to
neutralise the charge belonging to the electron or
electrons which undoubtedly exist in connection with
each atom.

(2) Or the bulk of the atom may consist of a
multitude of positive and negative electrons, interleaved, as it were, and holding themselves together
in a cluster by their mutual attractions, either in a
state of intricate orbital motion, or in some static
geometrical configuration, kept permanent by appropriate connexions.

(3) Or the bulk of the atom may be composed of
an indivisible unit of positive electricity, constituting
a presumably spherical mass or "jelly," in the midst
of which an electrically equivalent number of point
electrons are as it were 'sown'; these electrons
probably distributing themselves in rings, after the
fashion of Alfred Mayer's floating magnetic needles,
and revolving in regular orbits about the centre of

the jelly, with a force directed to that centre, and varying as the direct distance from it.

This hypothesis, in spite of obvious weaknesses connected with the nature of the positive unit, has great attractiveness :—for it explains the constant period of an orbit; it can explain the occurrence of visible radiation by perturbations of the orbit during collision; and it has been shown by J. J. Thomson to be capable of carrying us a long way towards a rational electrical theory of Mendeléeff's series of the chemical elements, together with some of their chemical—especially their electro-chemical—properties, and some features of their spectra. Moreover it goes further and explains in a fairly natural manner, and without artificiality, the gradual degradation of atomic energy by slow uncompensated and unperceived radiation; the consequent gradual oncoming of instability; and the occasional cataclysmic transmutation of one element into another, or rather into others, with explosive violence,—as observed in the facts of radio-activity. It gives in fact a rational though preliminary view of the hypothetical evolution of all matter, which many known circumstances now tend to support; and it accounts in a kinetic fashion for the immense store of intra-atomic energy. Nevertheless it is very far from being an established theory, and another view that can be taken of the rest of the atom is :

(4) That it consists of a kind of interlocked admixture of positive and negative electricity, indivisible and inseparable into units, and incapable of being appreciably sheared by applied forces, but incorporated together as a continuous mass; in the midst of which one or more isolated and individualised electrons may move about and carry on

that display of external activity which confers upon the atom its observed properties.

(5) A fifth view of the atom would regard it as a central 'sun,' of extremely concentrated positive electricity at the centre, with a multitude of electrons revolving in astronomical orbits, like asteroids, within its range of attraction. But this would give a law of inverse square for the force, and consequently periodic times dependent on distance, which appears not to correspond with anything satisfactorily observed.

All these views however are painfully indefinite, except the third one, which regards positive electricity as an indivisible unit of perfectly unresisting uniform material (though 'material' is not the right word), of spherical form the size of an atom, in the midst of which a definite geometrical arrangement of electrons are revolving with known frequency in specified groups or rings. The amount of outstanding vagueness in this view is obvious; and is necessary, so long as we know little or nothing about the intrinsic nature of what we experience as positive electricity; but at the same time all the rest of this hypothesis is definite enough, and enables mechanical laws and calculations to be applied with considerable fulness to the elucidation of the phenomena that would be displayed by such a 'model' or hypothetical combination. And if the so-calculated phenomena are found to correspond with fact, it assuredly lends some strength to the hypothetical basis on which the calculation is founded; although it is certain to have to be modified somewhat in the light of growing experience as discovery proceeds.

Whatever it may be worth, this is the only theory of the nature of the atom which has been to any great extent elaborated; and, extremely imperfect though it is at present, it is worthy of some attention from its own intrinsic interest. It will be found developed by Professor J. J. Thomson in the *Philosophical Magazine* for Dec. 1903 and March 1904; and a general idea of its main features can be gathered from his American "Silliman Lectures," published in 1904 by Constable as a book called *Electricity and Matter.*

Were it less hypothetical a further account of it would be given here, but an extremely recent paper by the same great Physicist has tended to reduce the whole subject to a state of exaggerated uncertainty; since he gives reasons, which appear to be sound ones, in the *Philosophical Magazine* for June 1906, for assuming that only one active electron is contained in a hydrogen atom, and that all other elements contain a number of electrons comparable to their atomic weight, reckoned on the basis that $H = 1$ (see Appendix L). This is an extraordinary and unsuspected result, and at first sight appears very unlikely, since the ordinary chemical assumption of a unit atomic weight for Hydrogen has always been known to be a pure convention, made for convenience alone, and not likely to correspond with anything in nature. I do not suppose that anyone imagined that it would, even provisionally, be found to have a physical and rational basis. The subject is further referred to in Chapters XVII. and XIX. below.

In that state of uncertainty the matter must be left for the present; but we may go on to indicate roughly how some of the known properties of matter could be expected or explained, on a view of the electrical constitution of matter which supposes it composed of a large number of positive and negative electric charges, irrespective of the particular mode of their aggregation and distribution.

CHAPTER XVI.

Nature of Cohesion.

WE shall now try to trace some of the consequences of the view that all atoms of matter are built up of the same fundamental units, and are composed of aggregates of a definite number of variously grouped negative and positive charges,—which for present purposes we may call electrons, even though some are positive—arranged in kinetic patterns and keeping apart by reason of the vigour of their own orbital motions.

At first it is not easy to do more than imagine the electrons to be statically grouped into regular patterns. It is easy to conceive this on the hypothesis No. 3 of last chapter; for though in free space they would be unstable or disperse, their possible groupings are easily calculated in a positive menstruum; for instance they might be arranged in triangular or square or hexagonal order; with other allied three-dimensional possibilities familiar to students of crystallography. See, for instance, William Barlow, *Brit. Assoc. Report*, 1896, p. 731; also Lord Kelvin, *Phil. Mag.*, March 1902, and elsewhere.

On Chemical and Molecular Forces.

The force of chemical affinity has long been known to be electrical. This opinion was propounded by Berzelius, and was also held previously by Davy and afterwards by Faraday. Ordinary electrical attraction between charged bodies may be called molar chemical action; only there is no combination in ordinary cases, because the opposing charges spark into one another, and so the attraction ceases when a certain proximity is reached. This discharge and cessation of attraction does not seem to occur among atoms; the difference of potential between them is too low to permit of mutual exchange or neutralisation of charge, so the combination is permanent.*

Real chemical attraction occurs between two atoms each of which contains an unbalanced electron—one extra, or it may be more than one extra, electron of given sign. Such an atom thus has a centre of force whereby it can attach itself to other atoms and enter into pairing or chemical combination with them. It is probable that a negative charge is an excess, and a positive charge a defect; and that when pairing occurs the excess charge of one fills up the deficiency of the other, and composes a complete and neutral molecule.

Union of this kind, however, never seems quite as strong and permanent as the union of the electrons in the atom itself: the molecule easily separates at the same place again, under the influence of decomposing influences, and does not seem able to split up in other ways into new substances; except in organic chemistry, where various modes of splitting up a

* See Lodge, *Brit. Assoc. Report*, 1885, pp. 744, 5.

complex molecule can be brought about, and are practically utilised for the generation of new compounds, *e.g.*—

$$CH_3 . CH_3 = C_2H_5 . H.$$

It is probable that the same sort of thing is *possible* with simple bodies, but that the so-called " elements " constitute a peculiarly stable group, the ingredients of which so far have only partially been re-associated into isomeric or allotropic forms, and have not yet been detached from each other.

When chemical combination occurs between two oppositely charged atoms, there is no electric discharge between them : the two atoms retain each its own charge, and cling together for that reason. When they are separated, each is an ion and possesses its appropriate charge.

It is possible to charge an assemblage of neutral molecules with an excess or with a defect of one or more electrons, by processes of ordinary electrification, such as friction ; but the attachment of these supernumerary electrons is loose—and they can be shaken away by the agitation of ultra-violet light and in many other ways. Even splashing of water into spray shakes some loose, and can thus perturb an electroscope, although the liquid was not charged beforehand ; * a fact which adds to the probability that the water unit is a molecular aggregate. And in the case of massive atoms, of high atomic weight, they occasionally appear automatically to reach a condition of instability, and rearrange themselves in such a way as to throw off one or more electrons spontaneously, which then fly off tangentially with whatever orbital velocity they may have had, giving

* Lenard on electrification near waterfalls. See also Chap. VII., above, on ionisation.

rise to part of phenomena recently discovered under the name of *radio-activity*. But instead of supposing that their violence of ejection is due to velocity previously possessed by them, it is possible to suppose that they are driven away by intrinsic static force, so that their previous energy was potential; and this is the form of hypothesis favoured by Lord Kelvin. See *Phil. Mag.* for March 1902.

Molecular Forces, Cohesion.

But there is another kind of adhesion or cohesion of molecules, not chemical but what is called molecular. This occurs between atoms not possessing ionic or extra charges, but each quite neutral, consisting of paired-off groups of electrons. At any moderate distance the force of attraction between paired electrons will be next to nothing, but at very minute distances it may be very great; ultimately becoming almost indistinguishable from chemical combination, except that the cling will be a weak cling at a multitude of points instead of an intense cling at only one.

Consider the outer surface of an atom consisting of a regular group of interleaved electrons of alternately opposite sign. Its equipotential surfaces will be dimpled or corrugated or pimply sheets, which at a little distance away will be almost plain; but the dimples will increase rapidly in depth and become like the cover of a mattrass, when something less than molecular distance—something approaching the internal electron distances apart—is reached.

Two such atoms will therefore tend to settle down with their equipotential surfaces adjusted into uniformity, the pimples of the one fitting into the hollows of the other; and this is the state of things suggested

by the facts of cohesion.   For a diagram representing
the state of things intended, see fig. 20.

To investigate the actual law of force would be
difficult, and too many assumptions would have to be
made for the geometrical arrangement of the electrons
in the adjacent atoms; it could only be approximate,
because we should probably, at least in the first
instance, have to assume a static distribution.
Nevertheless the attempt might be instructive, and
might in a developed form be suitable for an Adams
Prize Essay.

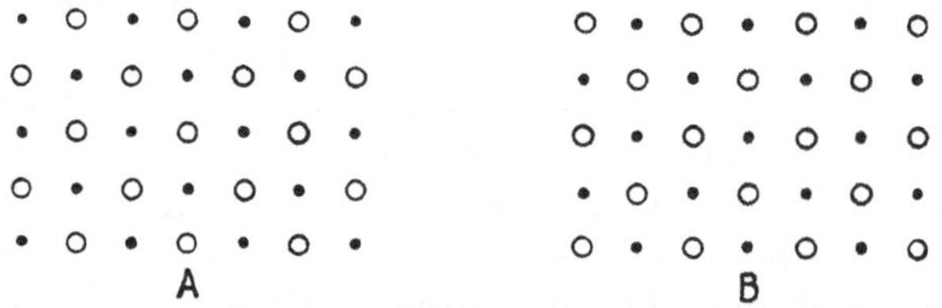

FIG. 20.—Ordinary Cohesion between two Neutral Atoms A and B:
each atom supposed to consist of interleaved electrons of opposite sign—
depicted in any convenient way—which attract each other by residual
or spare affinity.   This is due to a few of the lines of force which stretch
across the interspace, and hold the pair of atoms together.   The maximum distortional shear permissible depends on the ratio of the electronic
to the atomic distance.

It is quite plain, however, that the result would be
a force rapidly increasing and becoming great at
small distances, and practically nil at any perceptible
distance.

A theory of cohesion cannot really be given until
the structure of an atom is better known, but in
all probability it will proceed on lines not wholly
unlike the above.

Molecular forces on this view are electrical, just as
much electrical as are chemical forces; but they occur
between chemically saturated molecules, and are due
to the interaction or distant influence of paired

electrons on each other across molecular distances. It may be said to be a result of "residual affinity."*

Ions cannot thus combine; because if they were oppositely charged their combination would be chemical, and if they were similarly charged they would strongly repel each other. But if ions arrive at a metallic electrode, or are provided with other means of passing on their free charges, they cease to be ions; and then the diselectrified atoms can and do combine molecularly with each other.

It is of course possible for an ion to have more than one free electron, forming a dyad or a triad radical; and the way in which a neutral group can receive, and by rapid re-adjustment pass on, an extra foreign electron, reminding one of the re-adjustment of the films in a lather when one compartment bursts, is doubtless instructive.

The effect of electric polarisation on such a neutral group of electrons is noteworthy. The effect of a charged body in the neighbourhood is at once to disturb the equilibrium, and to perturb the grouping throughout the atom, more or less: it will cause the negative electrons to protrude slightly on one side and the positive on the other (see fig. 21 where two different but very complete kinds of polarisation are shown).

If two molecules were beyond each other's molecular range, and if the neighbouring surfaces could by any means—as by the supply of electricity from without —be oppositely electrified, the forces of cohesion would be intensified momentarily, by something akin to chemical affinity, and cohesion would set in over ultra-molecular distances. This appears to be what goes on in a "coherer." The opposite charges

* See Lodge in *Nature*, 1904, vol. 70, p. 176.

cannot be *maintained* electrostatically between two neighbouring metallic surfaces, but they can be momentarily imparted, by a sudden jerk or disruptive discharge or received electric impulse; and these are the things which are effective in promoting cohesion.

In the two diagrams;—fig. 20 represents a couple of atoms with interleaved electrons of opposite sign

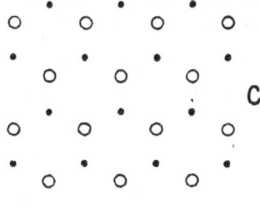

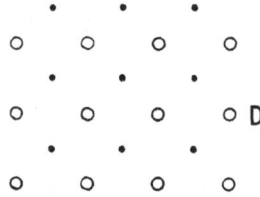

Fig. 21.—Two Polarised Atoms, illustrating electrically intensified cohesion.

in square order, the atoms being within range of one another and so cohering by molecular or non-chemical forces. They have adjusted themselves into a cohering position; but a vertical shear through half the distance apart of the electrons would disintegrate them. An angle represented by half the electron-distance divided by the molecular distance, is therefore a measure of the maximum distortion a substance can undergo.

Fig. 21 shows a couple of atoms both electrically polarised, as by a positively charged rod held above both. To vary the illustration, the constituents of C are shown polarised into hexagonal order—an effect such as might also be caused by lateral pressure in some cases; while the constituents of D are depicted in diagonal square order—which has the effect of violent electric polarisation. In any case polarised atoms such as C and D are clinging by forces much stronger than the force of ordinary cohesion at that distance. They represent adjacent atoms of a momentarily polarised coherer.

It is not to be supposed that the electrons in a polarised atom need really ever be disturbed as much as is shown in the diagram, nor any more than an almost imperceptible amount, in order to produce this chemical cohesion effect. For that is what polarisation accomplishes : it converts ordinary molecular force, or cohesion, into incipient but real chemical affinity ; both kinds of forces being on the above hypothesis electrically explicable.

CHAPTER XVII.

FURTHER CONSIDERATIONS REGARDING THE STRUCTURE OF AN ATOM.

THE hypothesis that has been already referred to, No. 3 in Chap. XV., that an atom consists of a globular mass of positive electricity with minute negative electrons embedded in it—either at rest, or vibrating about a position of equilibrium, or revolving in regular annular orbits,—is obviously not free from difficulties; though its success in explaining many observed facts seems to justify an attempt to minimise those difficulties, and to render hopeful an effort ultimately to overcome them.

One objection that can easily be raised is to ask how a mass of positive electricity can hold together against the mutual repulsion of its parts. This difficulty is felt more in the case of positive electricity than in the case of negative, because by hypothesis the positive charge has some perceptible bulk, namely the size of an atom, whereas the negative charge or electron is exceedingly small. But however small an electron is, it must be supposed to have parts, and there is just as much reason for supposing the parts of a negative unit to be mutually repulsive as there is in supposing the parts of the positive unit to be mutually repulsive. Hence the question is equally

valid, how do the parts of an electron hold together?
But nobody seems to ask this question; and I see no
reason why it should be asked, because there is no
evidence that the parts of an electron *are* mutually
repulsive—there is no real evidence as yet that it
has parts at all.   What we know is that different
electrons repel each other, and are attracted by
positive charges, but we do not know that the parts
of one electron repel each other; in fact we know
nothing about the parts of an electron.   The lines
of force, or field of force, representing the activity
of an electron, may be entirely outside itself, and
need not penetrate into its interior at all: it may
be for all electrical purposes an indivisible unit.

But then, supposing all this admitted for a negative
charge, why should it not be extended to cover a
positive charge also?   Why should the parts of a
positive unit be mutually repulsive?   It is no answer
to say that the unit is bigger, unless the electron is
thought of as a geometrical point; the argument
about parts is just as valid in the one case as in
the other, and no more so.   What we require is some
conception as to the nature of the positive charge,
—and that I confess is wanting; though "an
entanglement of finite size" seems to Larmor quite
possible and natural; and anyhow an argument
against its existence based upon the assumed repul-
siveness of its parts does not seem to be an argument
of any weight.

Another suggestion can be made: and that is that
the main bulk of the atom, in which electrons are
embedded, consists not of positive electricity alone
but of a close admixture or combination of positive
and negative electricities—inseparable and as it were
rigidly connected—behaving to outside forces like the

L. E.   L

oxygen and hydrogen interlocked together in a gallon of water ; in which water a few extra atoms of oxygen might easily be dissolved, in a relatively free condition—that is to say, not forming part of the fixed constituent-oxygen—just as a few outstanding electrons may exist in the general mass of an atom.

This is the hypothesis numbered 4 in that list of alternative schemes which was exhibited in Chap. XV.; and the latest results of Thomson, there also mentioned, tend towards supporting it.

If any lines of force are then postulated between the constituent positive and negative fluids inside an atom they would be entirely of an internal character, and only those belonging to some free unbalanced charges would emerge and produce ionic activities : the internal lines of force would be wholly occupied in holding the atom together, and would have no influence on neighbouring atoms—not even molecular or residual influence. The main bulk of atom in that case would be like a finely interleaved condenser, incapable of discharge or of decomposition,—except by actual disintegration such as might occasionally accompany or cause radio-activity.

CHAPTER XVIII.

SUMMARY OF OTHER CONSEQUENCES OF ELECTRON THEORY.

Radio-activity.

IF many atoms of a substance have electrons attached to them, and if these are performing orbital revolutions, it is natural to ask how then can it be that substances are not constantly emitting waves and radiating away their energy. For we have seen in Chap. X. that electric charges in revolution or vibration constitute radiators, and must emit more or less radiation; thereby dissipating their kinetic energy and gradually either coming to rest or effecting some other change. Fortunately, owing to the brilliant researches of Becquerel, Curie, and others, certain substances have been found in which the radiation intensity reaches a very perceptible magnitude; and it appears that this radiation may be of several kinds—

    1st, of waves or pulses analogous to Röntgen radiation : called γ rays ;

    2nd, of rays analogous to cathode rays consisting of electrons bodily shot off: called β rays ;

    3rd, of positively charged ions or atoms, or

semi-atoms, consisting of something like helium apparently, likewise shot off with great energy : called α rays ;

4th, as a consequence of all this radiation, detached portions of the residue of the substance drift away, not charged with electricity, but emanating something after the fashion of an odour. This gaseous emanation is found itself to possess a very high intrinsic radiating power, and to be capable of attaching itself to, or causing a deposit on, other materials in the neighbourhood, so that they too acquire temporary radiating power : a deposit at one time spoken of as induced or excited activity.*

The substances which possess any noteworthy amount of this radiating power are substances with very high atomic weight; and their emitting power would appear to be probably due to an internal commotion or convulsion, of sufficient violence to detach and expel, every now and then, some particle or fragment ; and also, by the shock of the expulsion, to generate some feeble but exceedingly penetrating Röntgen rays.

---

* See, for instance, papers by the original discoverer of spontaneous radio-activity, M. Henri Becquerel, in *Comptes Rendus*, 1896 and 1897 ; see also Rutherford, *Phil. Mag.*, January, 1899 and 1900, with quantitative determinations concerning it. Also in *Phil. Mag.*, July and November, 1902. Other references are M. and Mme. Curie, *Comptes Rendus*, November, 1899 ; Hon. R. J. Strutt, *Phil. Trans.*, A 1901, vol. 196, p. 525 ; Sir W. Crookes, *Proc. Roy. Soc.*, vol. 66, p. 409 (1900), vol. 69, p. 413 (1902), also "Electrical Evaporation," 1891, *Proc. Roy. Soc.*, vol. 50, p. 88 ; and many other workers. References to them are now conveniently collected in Professor Rutherford's excellent treatise.

Madame Curie's original Thesis on Radio-activity for her Doctorate, of date 1903, is a masterly production.

It is easy to grant that, whenever there are actual collisions of sufficient suddenness, some radiation must be emitted; but we cannot help asking, why does not the quiet orbital revolution of electrons round atoms, in a substance not in a high state of thermal disturbance and not possessing specially massive atoms, why does not this also give rise to a perceptible amount of true radiation and loss of energy? One answer that has been given is as follows:—

The radiators are not isolated or independent, and surface radiation is maintained by layers at greater depth in the substance. Moreover the radiators are so close together that they are in all sorts of phases within the first quarter wave length, a length which embraces a multitude of them; wherefore a multitude is a worse radiator than one, because they interfere and produce but little external or distant effect; like the two prongs of a fork, or two neighbouring organ pipes, or the front and back of a vibrating wire. See Larmor, *Ether and Matter*, page 232; the proposition by which he considers the question settled is that the vector sum of accelerations equals zero.

Of course it must not be forgotten that radiation of a low temperature order is as a matter of fact always going on from all substances; that energy is conserved, and constancy of temperature persists, merely because loss is equal to gain, because absorption compensates radiation, not because radiation ceases; and that to make an estimate of the amount of radiation, so occurring, it would be necessary to suppose the body in an enclosure at absolute zero: when undoubtedly its kinetic energy *would* rapidly leak away, and be dissipated. But this refers to questions connected with ordinary radiation, whereas we are now dealing with the extraordinary variety known as radio-activity.

If radiation goes on at the expense of the internal energy of an atom, as it must if the atom contains revolving electrons,—still more if, as on the electric theory of matter, it is wholly composed of electric charges,—it becomes necessary to ask further :—Why are not all atoms temporary and unstable ? Why are they not all liable to internal catastrophe and disruption, akin to earthquakes and volcanoes ? Why do they not all exhibit the phenomena of radio-activity ?

The whole subject of radio-activity is a large one, upon which I do not propose to enter at any length here. Suffice it to realise that any difficulty of explanation, in connection with it, is not the fact itself, but rather the question why it is not more notorious.

However, so far as that most striking and interesting phenomenon—the excessive photographic and electric radio-activity of certain rare substances—is concerned, it has been already hinted that the greater part of that does not consist so much in the emission of radiation proper—whether in the form of pulses of X-rays or any other form—as in the flinging off of particles ; sometimes negatively charged particles or electrons, sometimes positive ions. And these expelled particles, when they strike a photographic plate, appear to generate by the concussion electric waves which affect the silver salt. The faint photographic influence of ordinary substances, observed by Dr. W. H. Russell, seemed to suggest that incipient power of this kind is not limited to bodies with heavy atoms, like Uranium, Radium, Polonium, etc., as described by Becquerel and the Curies, though these substances show it to an extraordinary degree : Dr. Russell, however, appears to have traced his at first interesting effects to the merely chemical action of hydrogen peroxide. He has quite recently shown that leaves of growing plants have a spontaneous photographic power of the same kind.

The whole subject, together with the allied one of the loss of charge from hot bodies,* first discovered by Dr. Guthrie long ago (see *Phil. Mag.*, [4], xlvi. p. 273), is one that demands special attention and treatment.

Crookes discovered that the alpha rays, or particles projected by radium, polonium, and other strongly radio-active substances, were able by their concussion with a target of zinc-sulphide to produce luminous flashes, visible under slight magnification : and the evidence on the whole is in favour of the view that many of the impinging atoms may have speed enough to be able to cause a separate flash, by its own individual action on the crystalline compound : a phenomenon popularised in the little pocket-instrument called a ' spinthariscope.' When we consider the speed with which these particles are ejected, such an idea is not surprising, although it is novel to have to contemplate any perceptible effect produced by a single atom. But taking these projectiles to be atoms of hydrogen or helium, of mass 10^{-24} grammes, flying with say one-tenth the speed of light,—the stoppage of one of them within molecular dimension, that is within its own thickness as ordinarily estimated, 10^{-8} centimetre, would require, for an exceedingly minute fraction of time, the expenditure of 80 horse-power.

* See, for instance, Strutt on leakage from hot bodies, *Phil. Mag.*, July, 1902 ; J. J. Thomson, ditto, *Phil. Mag.*, August, 1902 ; further developed by O. W. Richardson, *Phil. Trans.*, vol. 201, p. 516 (1903).

Solar Corona, Magnetic Storms, and Auroræ.

Another subject on which it is tempting to enlarge is the explanation of various astronomical and meteorological phenomena by the electron theory. The theory of Auroræ has recently been elaborated by Arrhenius; but the whole doctrine of emanations from the sun, and of repulsion of small particles both by his light and by his probable electrification, is a matter that has been familiar to me for many years, through conversation with FitzGerald and others. See, for instance, Larmor, *Phil. Trans.*, 1894, vol. 185, p. 813; Lodge on Sunspots, Magnetic Storms, Comets' Tails, Atmospheric Electricity, and Auroræ, in the *Electrician* for December 7, 1900, vol. 46, p. 250; FitzGerald, *Electrician*, December 14, 1900, with reference to his review of a Heaviside volume in 1893 (*Electrician*, vol. 31, p. 390). See also Fitz-Gerald's collected "Scientific Writings," at date 1882.

The earth is in fact a target exposed to cathode rays, or rather to electrons, emitted by a hot body, viz. the sun. The sun is evidently intensely radio-active : and the result of its discharge of electrons into the approximate vacuum of its immediate neighbourhood is not unlikely to be the appearance known as the corona. The gradual accumulation of negative electricity by the earth is a natural consequence of this electron bombardment extending to greater distances across space, where no residual matter exists ; and the fact that the torrent of particles constitutes an electric current of fair strength gives an easy explanation of one class of magnetic storms ; these storms having long been known, by the method of concomitant variations, to be connected with sun-spots and auroræ. The electric nuclei, when they

form ions, would also serve as centres for condensation
of atmospheric water vapour at high altitudes, and so
be liable to affect rainfall.   Moreover, the fact that
water vapour condenses more readily on negative
than on positive ions, seems to furnish us with one
explanation of atmospheric electricity; for a fall of
rain would bring down with it a negative charge,
and would leave the upper regions positively
electrified with respect to the earth's surface : and
this agrees with the known sign of the normal field
of electric force in the atmosphere.

These early perceptions have been well elaborated
of late by Arrhenius; and his explanation of the
aurora—-by means of the catching and guiding of
rapidly moving electrons by the earth's magnetic
lines of force, so as to deflect them away from the
tropical sunshine, and to guide them in long spirals,
along the lines, to the poles,—there to reproduce the
phenomena of the vacuum-tube in the rarified upper
regions of the atmosphere—is particularly definite and
pleasing.   Some of the other astronomical suggestions
he has made are likewise of considerable interest.

Transformations of Radium, etc.

The following details are given by Rutherford for
the spontaneous transformations which radium under-
goes, together with the lifetime constant of the
various products—that is to say the time required
for the activity of the product to fall to one-half its
value,—and also the kind of particles or rays which
are thrown off.

Certain of the changes are rayless changes, and
may be considered to result in an allotropic modification, without change of atomic weight; but
whenever an alpha particle is thrown off, the atomic

weight must change, presumably by an amount appropriate to the loss of an atom of helium. It will be observed that the gamma or Röntgen rays always accompany the emission of a beta particle or electron, and never appear otherwise ; also that it is only in some of the changes that electrons are thrown off. It will be understood that the so-called "emanations" are all of the nature of gas, while the other products are like a solid deposit or coating. These active deposits, or "excited activities," can be subjected to ordinary chemical tests : some of them are soluble in acids, some in ammonia ; some of them are volatile at a high temperature, some are not readily volatile. The following table contains the principal features of the transformations of the more remarkable radio-active substances ; for more details Rutherford's book must be referred to. The product here called Radium F appears to be the same as that called by other observers Radio-tellurium, or by its original discoverer, Mme. Curie, Polonium :

The substance	has a life-constant	it shoots off	and changes into
Uranium	(say) 600 million years	an α particle or (?) helium atom	Ur. X
Ur. X	22 days	$\beta\gamma$ rays	
Thorium	24×10^9 years	an α particle	Th. X
Th. X	4 days	α particle	Emanation
Emanation	54 secs.	α particle	Th. A
Th. A	11 hours	no rays	Th. B
Th. B	55 mins.	$\alpha\beta\gamma$ rays	
Actinium		no rays	Act. X
Act. X	10 days	α particle	Emanation
Emanation	4 secs.	α particle	Act. A
Act. A	36 mins.	no rays	Act. B
Act. B	2 mins.	$\alpha\beta\gamma$ rays	

The substance	has a life-constant	it shoots off	and changes into
Radium	1300 years	α particle	Emanation
Emanation	4 days	α particle	Radium A
Radium A	3 mins.	α particle	Radium B
Radium B	21 mins.	no rays	Radium C
Radium C	28 mins.	αβγ rays	Radium D
Radium D	40 years	no rays	Radium E
Radium E	6 days	βγ rays	Radium F
Radium F	143 days	α particle	Possibly Lead

Emanations.

The discovery of thorium and radium emanations was made by Rutherford and Dorn respectively in consequence of an observation of Owens on the irregularity of thorium rays in producing ionisation, the fact being that any of these materials are more active when the emanation has been allowed to accumulate than soon after it has been removed. For the emanation, although so infinitesimal in quantity, is considerably more active than the substance itself; and, being a gas, it can readily be drawn away or otherwise expelled from the pores or neighbourhood of the salt. But it accumulates again, being evidently generated *in situ*, and presently the full activity of the substance is restored. Radium emanation is shown by Rutherford and Soddy to liquefy at a temperature of about 150 degrees below zero; thorium emanation liquefies at about $-120°$ C. They appear to be quite definite, though transitory and very unstable and disintegrating, materials.

Deflexion of Alpha-rays.

When alpha-rays are submitted to a strong magnetic field they are deflected, though very

slightly, in a direction indicating that they are positively charged particles. Rutherford and Becquerel
have both observed this fact, and Strutt has confirmed
the fact of positive charge. But quite recently *
Soddy has surmised that this positive charge might
be acquired by ionisation in travelling through the
air, and that in a high enough vacuum no deflexion
would be observed ; thereby showing that intrinsically they did not possess a charge. He believes
himself to have confirmed this by careful though
difficult experiment. So important, and in some
respects improbable, a conclusion, however, cannot
yet be regarded as at all certain.

Rutherford was the first to observe the fact and
the sign of the magnetic deflexion or curvature of
alpha rays, and to make an estimate of its amount.
He invented the device of sending them through a
magnetic field, up a stream of rarified hydrogen,
between a set of narrow plates set edgeways ; which
latter constituted a grid that would be opaque
unless the trajectories of the flying particles were
rigorously straight. He thus made the discovery that
the rays consisted of positively charged particles, and
arrived at a rough estimate of their atomic weight.

Becquerel measured the magnetic deflexion of
alpha rays, at different distances from the source,
by letting them graze a photographic plate at a
known angle. The actual trace observed was a
short slant line with a slight curvature ; reversal of
the field slanted the line the other way, thus giving
a resultant impression like the conventional two wings
of a flying bird drawn at a very acute angle. Subsequent measurement of the distance apart of different
positions of the two wings gave the data sought.

* See *Nature*, 2nd August, 1906.

Rutherford has examined the deflexion of alpha particles from radium in a careful manner. Previous experiments, such as his own and those of Becquerel and Des Coudres, were made on a thick layer of radium ; but under these circumstances the particles are projected with a considerable range of velocity. To obtain homogeneous radiation it is necessary to use a very thin layer ; and Rutherford employed for this purpose the product Radium C,—namely, part of the active deposit which appears on a fine wire exposed for some hours to radium emanation. This is a deposit of utterly imperceptible thickness, undetectable by any means save radio-activity ; and it consists of Radium A, B, and C. The activity of Radium A disappears in about a quarter of an hour ; Radium B emits no rays ; so Radium C alone is left active, and it emits only alpha rays. It is true it dies away in about a couple of hours, but there is time enough for an experiment. The method employed is like this :

Rays from the wire pass through a narrow slit, and then on to a photographic plate in a vacuum ; a uniform magnetic field is applied in a direction parallel to the slit, so as to curve the rays ; this field is reversed every ten minutes, so that on developing the plate two narrow parallel lines are observed, the distance between which represents twice the deflexion ; their sharpness shows that the rays were homogeneous. The path corresponding to a magnetic field of 9470 c.g.s. units had a radius of curvature equal to 42 centimetres.

Electric deflexion was not then applied ; but, estimating the number of particles expelled from the Radium C as $6·2 \times 10^{10}$ per second, and the heating effect as 31 calories per hour, Rutherford calculated the emission of energy as ·36 million ergs per second. Combining this with the magnetic deflexion, the result is that $u = 2.6 \times 10^9$ centimetres per second, while $e/m = 6,500$ electro-magnetic units. Des Coudres, using both electric and magnetic deflexion, but employing more complex rays, obtained $u = 1.65 \times 10^9$, and e/m 6,400.

Activity of Radium at all Temperatures.

Messrs. Dewar and Curie have shown that at the temperature of liquid hydrogen, 252 degrees below zero centigrade, the heat evolution of radium is the same as at ordinary temperatures. It has also been shown to be equally active in the solid or in the dissolved state, and elevation of temperature makes it no more active.

The activity of radium, as observed in Crookes' spinthariscope, is just as marked at the temperature of liquid hydrogen as at ordinary temperatures, provided the luminescent screen is not likewise chilled.

Spectrum of Radium.

In speaking of the spectrum of radium there is an ambiguity to be guarded against. We may mean the spectrum of radium itself, when it is put in a flame or subjected to the electric spark; or we may mean the spectrum of the spontaneous glow of radium chloride or bromide, which can be seen in the dark. The latter spectrum is very difficult to observe, because of its extreme faintness, but the experience of Sir William and Lady Huggins has enabled them to examine it, and to show that it is chiefly the spectrum of atmospheric nitrogen; which is thereby proved to be ionised and violently disturbed by the neighbourhood of radium, to the extent of emitting luminous radiation without elevation of temperature. The spontaneous radium-glow is, in fact, probably due to its influence on other substances ; and ordinary glass, exposed to it, darkens and becomes thermo-luminescent,—that is to say, it begins to emit light when raised to a temperature of about 500 degrees.

M. Eugène Néculcea gives the following table of the photographic spectrum ·of the radium spark, as observed by M. Demarçay, the wave lengths being given in "tenth-metres." In the visible spectrum, which is not photographed, there is only one notable ray, of wave-length 5665. For other observations, see Runge, *Astrophysical Journal*, 1900, p. 1.

WAVE-LENGTH.		INTENSITY.	WAVE-LENGTH.		INTENSITY.
Blue	4826·3	10	Violet	4600·3	3
	4726·9	5		4533·5	9
	4699·8	3		4436·1	8
	4692·1	7	Ultra-violet	4340·6	12
	4683·0	14		3814·7	16
	4641·9	4		3649·6	12

The detection of radium by the spectroscope, through its strongest line, 3814·7, though it may be a method perhaps a million times more sensitive than ordinary chemical analysis, has been shown to be a million times less sensitive than a method of detection by means of an electroscope, utilising the extraordinary ionising power of its radio-activity. For Rutherford reckons that each α particle expelled from radium is able to generate some hundred thousand ions before it is stopped, or rather before its immense initial velocity falls below a certain critical speed at which ionisation ceases to be caused. A charged electroscope of proper sensitiveness is able to indicate by its leak the presence of this number of ions, and accordingly is able to signal the presence of only one *a* particle. But the number of such particles expelled by a milligramme of fully active radium is estimated as a hundred million every second.

Electric Production.

The power of radium constantly to develop electricity was first demonstrated by W. Wien, by suspending a tube in a vacuum. Dorn observed that sometimes sparks were obtained on opening a tube containing radium. Strutt contrived an ingenious device for displaying this constant production of electricity by reason of the very different penetrating power of the $+ a$ and the $- \beta$ particles, so that one set can be trapped while the other set escapes.

Fig. 22 shows Strutt's apparatus and "perpetual" —more accurately, perennial—electric generator and source of mechanical energy. A small tube containing radium salt, with its outside made sufficiently conducting, has a pair of gold or aluminium leaves attached to it, and is insulated in a very high vacuum by a quartz rod; a metallic strip, lining the vacuum case, being connected to earth. The radium fires off positive and negative particles in equal quantities; but the negative, being small and penetrating, are many of them able to escape, while the positive accumulate and thus keep on slowly charging the gold leaves with positive electricity till they periodically overflow. The high vacuum is necessary to prevent the internal atmosphere from becoming conducting, by ionisation from the shock of the flying particles, and as a consequence preventing the gold leaves from becoming charged.

Radio-activity of Ordinary Materials.

Even in the absence of radium, a charged electroscope is usually found to leak rather more than any creeping along the supports can account for; this

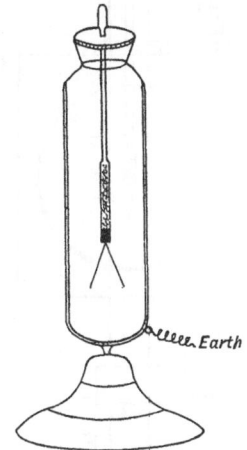

FIG. 22.—Strutt's mode of demonstrating constant electric production by radium.

leakage must be due to ionisation of the atmosphere, and a consequent kind of gaseous electrolysis in the air. The ionisation is partly due to stray radiation from outside, but some of it may be due to the radio-activity of materials with which the interior of the electroscope can be lined. Definite experiments of this kind have been made by M'Lennan and Burton of Toronto, and by Strutt, also by A. Wood, Rutherford and Cooke. When cylinders of zinc, tin, or lead were used to line the electroscope, a certain leakage was observed, but this was reduced some thirty per cent. when the whole was plunged into a tank of water so as to screen the interior from outside radiation.

Rutherford and Cooke surrounded the electroscope with thick screens of various kinds, and once with

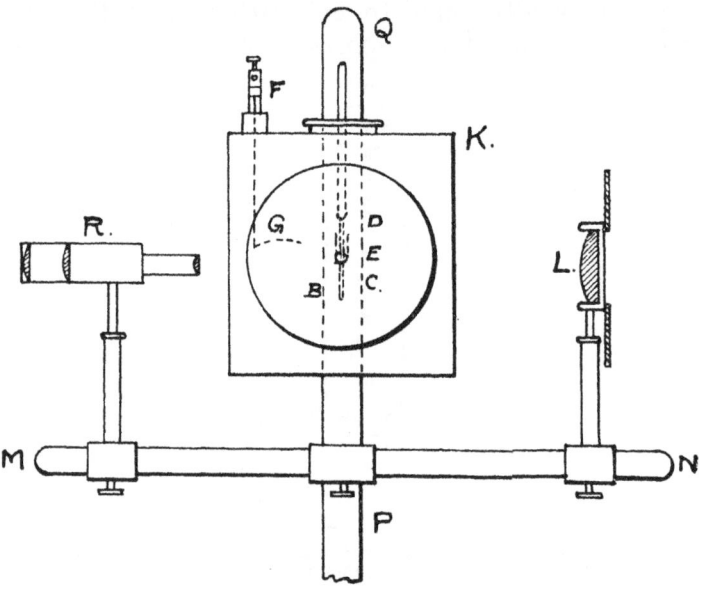

FIG. 23.—Righi form of electroscope or electrometer, of small capacity. The vital parts DECB, with the gold leaf, are shown on a larger scale in fig. 24. They are contained in a metal case, with a movable wire G, by which they may be charged from outside. The instrument is read by a reading microscope R; an ordinary millimetre scale being placed in a conjugate focus and projected by the lens L on to the plane of observation where the gold leaf is. The radio-active substances are presented to the box outside a window of very thin aluminium foil.

as much as five tons of lead. A moderate amount of screen, however, was found to produce the same effect as a greater amount, showing that the influence of the outside penetrating-radiation could be checked without stopping more than thirty per cent. of the leakage ; the remaining seventy per cent. seemed due to internal radio-activity. Materials such as brick seem specially radio-active, and any metals which have been exposed to the outside atmosphere are more radio-active than virgin metal.

But, in addition to this induced or excited activity, Strutt and others tried a good many materials and found different characteristic effects with each.

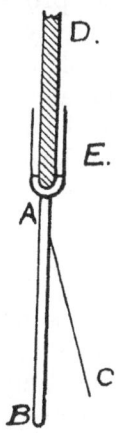

FIG. 24.—Working part of the above electroscope, magnified four times. The rod AB with the gold leaf C is cemented to a thin rod of melted quartz D by means of a drop of mastic or guttapercha at the bottom of a minute metallic capsule E attached to the rod AB ; thereby lessening any tendency of the charge to creep, and keeping the capacity exceedingly small.

The question of whether radiation is emitted spontaneously by metals of every kind, and not only by those few substances which are conspicuously radio-active, has been still further examined by Mr. Norman R. Campbell of Trinity College, Cambridge (*Phil. Mag.*, Feb. 7, 1906). He adduces very strong experimental proof that such radiating power actually exists, since they all ionize the air in their neighbourhood. Moreover, the experiments indicate a characteristic or specific quality in the radio-activity of the different substances: a result tending to negative the idea that it is due to some residual effect of a common impurity, such as an excessively minute trace of radium common to all metals; the indication is rather in favour of a specific radio-activity belonging to each metal, *not* due to any impurity.

It is not yet absolutely proved that this is identical with orthodox radio-activity, of the kind which is accompanied by atomic change or transformation of substance; but, inasmuch as the rays appear to consist to a great extent of alpha-rays, there is not much doubt but that the complete identity of the process will be established before long.

Population Analogy.

Since radio-activity is a sign of, and is accompanied by, disintegration and loss of material, it is manifest that substances of exceedingly high radio-activity must be comparatively scarce. Ordinary permanent materials cannot be violently radio-active, though each gramme of them might lose a few thousands of atoms per second without any probability of our being able to detect the loss by weighing—not even by weighings continued through a century. The plentifulness of a substance must depend on its rate of production, its life time, and its rate of decay; just as the population of a circumscribed area is determined by the birth rate, death rate, and average age.

CHAPTER XIX.

RADIATION FROM A RING OF ELECTRONS, AND ITS BEARING ON THE CONSTITUTION OF AN ATOM.

ALTHOUGH the radiating power of a single vibrating or revolving electron is considerable, even when the speed is not excessively high, it must be observed that the amount of radiation emitted is greatly diminished when a second symmetrically situated particle is introduced; because, at a distance, it will be virtually in opposite phase to the first. The diminution is specially marked when the speed is low. In order to assist the radiation, the second electron at the opposite end of a diameter should be of opposite sign to the first.

Three similar electrons at the corners of an equilateral triangle will radiate much less than two; and so with every addition to a symmetrical system of rotating particles, the radiating power diminishes; until when they form a continuous ring there is of course no radiation emitted at all, since everything is then continuous.

Professor J. J. Thomson has calculated the amount by which the radiation is diminished when the number of particles is increased: and shown how greatly it depends on their velocity. If they are travelling with an orbital velocity of one-tenth the speed of light, two electrons at opposite ends of a diameter radiate about one-tenth as much as either alone; four electrons at the corners of a square, likewise rotating with one-tenth of the velocity of light in its own plane, will have a radiating power of about one six-thousandth of that of one of them.

But if the velocity of the particles is only one-hundredth that of light—as must often, perhaps usually, be the case among constituent electrons in an atom,—then a pair would radiate only one-thousandth as much as one. For three, the radiating power is diminished to about one two-millionth; and for six, to something getting on for a trillionth,—which practically means no radiating power at all.

The actual numbers, and the calculation, can be found in the *Philosophical Magazine* for December 1903, p. 681.

All this depends on the electrons being symmetrically situated round the centre of rotation; but if by any cause—such as an atomic clash or chemical collision—they are displaced from symmetry, then their centre of gravity will rotate round the original centre, and will act as a single electron, or rather as an electron of multiple mass and constitution. The constituent particles will now not compensate each other at all, so far as this excentric motion is concerned. Accordingly the radiation instantly becomes violent, and must be regarded as the source of the visible spectrum: the nature of the lines depending on the structure of the composite body, which, by reason of temporary displacement, now acts as a single radiator of great power. In this way it is possible to conceive of the particular kind of radiation, exhibited in the spectrum, as being determined by the character of the composite radiator; while the occurrence of the radiation itself is limited to the periods during which the displacement by chemical clash or collision continues effective : a period in each individual case probably very short, but rapidly renewed in the aggregate by the combination or collision of other molecules.

Radiation due to high temperature may be caused in a somewhat similarly accidental and violent fashion, not occurring at all appreciably during the intervals of peace or regular motion.

Instability of an Atom.

The principle on which instability of an atom is to be expected, on the electrical theory, at certain critical points in its history, may, according to J. J. Thomson, depend upon the fact that a rotating ring of electrons is only stable so long as their angular velocity exceeds a certain critical value. Three electrons indeed are stable even when stationary, though they become more stable by rotation ; but four corpuscles in one plane, at the corners of a square, are not stable unless they are rotating with an angular velocity greater than $\sqrt{\left(\cdot 325 \dfrac{Ne^2}{\kappa m b^3} \right)}$, or $\cdot 57 c$ as it is called in the next chapter. Whenever their speed falls below that value, they adjust themselves at the corners of a tetrahedron or triangular pyramid,—tumbling over into the new position with a rapid collapse, analogous to the tumbling over of a top when its rate of spin has fallen below a certain critical value.

With a greater rate of rotation, five corpuscles in one plane may be stable ; but with six in a ring, stability is impossible at any speed, so long as there are only six present; but if one of the six, or if a seventh, be placed as a centre to the ring, then they become stable again ; and at a proper speed there can now be seven or even eight corpuscles in a ring, but not more, unless more corpuscles be placed in the centre. Suppose, for instance, that three be put in the centre : they will there form a triangle, and round them there may be a ring of as many as ten others.   But to get a stable ring of twelve corpuscles, you must have seven inside altogether ; of which six or five can be in another ring, with one or two at the centre of that. By this means a large number of corpuscles, all in rapid rotation, can be arranged in a series of rings ; but if the speed of any ring falls below a certain critical value it becomes unstable, and then there has to be a readjustment of corpuscles into another pattern, which must give rise to a sudden convulsion in the hypothetical structure of the atom. This readjustment involves a decrease of potential energy and consequent increase of kinetic energy, and hence might result in the expulsion of some corpuscles.

Note that such a convulsion, on this theory, must undoubtedly occur from time to time, though it may be a rare occurrence in the life of any one atom ; because revolving charges of electricity necessarily radiate energy to some extent—though usually to a very small extent,—and accordingly their speed must gradually be reduced, until sooner or later it arrives at the critical value at which a convulsion must occur. The convulsion is followed by readjustment into another pattern, and consequent transmutation into another element, or at least into an allotropic form of the first element if no fragment is lost ; and the occurrence is accompanied, and indeed demonstrated, by radio-activity.

A list of the series of changes actually observed in radio-active substances will be found above in Chap. XVIII., p. 170.

Cosmic Analogy.

An analogy may be drawn between the occasional disruption of one of a large group of atoms, and the phenomenon observed from time to time in the sky called a new or temporary star. Both are outbursts of a kind of radio-activity, though they may be excited by different causes ; both are comparatively rare occurrences, when the whole available number of bodies capable of such outburst or collision is contemplated.

Assuming that in a gramme of average terrestrial material there are a thousand such eruptions every second, that would correspond to about one new star per century among a cosmic assemblage of ten thousand million bodies.

Another Account of Atomic Instability.

A different idea of the nature of Instability was suggested by myself. An electron which is rotating outside an atom, being attracted thereto by a force varying as the inverse square of the distance, will, as it tends to lose energy by radiation, get drawn nearer and nearer to the atom ; and will increase in speed inversely with the square root of the distance. As the speed thus increases, the effective inertia of the particle will increase, and accordingly it will be more and more difficult to retain it in an orbit by the centripetal force ; since this force, being merely an electrical attraction, is not a function of the speed, or, in so far as it is a function of the speed, it is a function which diminishes as the speed increases. Accordingly there must come a time when the electrical attraction is incompetent to hold in the revolving mass, which then begins to strain itself a little further off, with the velocity now acquired; but, by so doing, it still further diminishes the force which holds it in, without diminishing the centrifugal force it is itself exerting; and accordingly there is no longer equilibrium. The equilibrium at the point of greatest proximity is in fact unstable, and so the particle itself flies off tangentially with the speed which it had then acquired, thus beginning a radio-activity of a fresh kind—radio-activity discovered by Becquerel—the emission of violently flying electrons or Beta-rays.

The whole constitution of the atom may be upset by losses of this kind, and a rearrangement of its substance appears occasionally to occur, with the flinging away of some portion as a material projectile; these particles thus thrown off constituting the observed Alpha-rays. Sometimes electrons are thrown off too.

The sudden ejection of an electron, like the sudden stoppage of one, is well calculated to excite those vibrations in the ether discovered by Röntgen—known in the case of spontaneous radio-activity as Gamma-rays.

Electric Theory of Matter.

A scheme or model for the construction of atoms of different sorts of elementary substances, by means of groups of electrons revolving in plane orbits round a centre according to the law of direct distance,—together with some indication of the known spectral

and chemical properties of such substances, and the
deduction of a series or family likeness akin to
Mendeléjeff's series on this hypothetical constitution,
—is to be found worked out in a masterly manner
by Professor J. J. Thomson in the *Philosophical
Magazine* for March, 1904.

If at the present time there were other confirmatory
evidence of this mode of regarding the atom—if it
were possible to understand what is meant by positive
electricity existing as a permeable sphere with a large
number of electrons sown in it—it would be proper to
give a further account of this remarkable and attractive
theory; but inasmuch as the light which has quite
recently fallen upon the whole subject is of a flickering and undecided nature, tending somewhat in the
direction of confusion and uncertainty rather than confirmation of this promising but still necessarily vague
hypothesis—see Chaps. XV. and XVII. and Chap.
XX.—I think it better to leave its further discussion
and development to a future time, and meanwhile
await further theoretical and experimental evidence.
For it is neither theory alone nor experiment alone
which can be of service in this matter : progress
must be achieved by a combination of the two,
welded together by genius; and it is just this
combination of theory and experiment—ingenious
experiment combined with advanced mathematical
theory—to which of late years all the finest advances
in physical science have been due.

CHAPTER XX.

DIFFICULTIES CONNECTED WITH THE ELECTRIC THEORY OF MATTER.

1. *Concerning the formation of Spectrum lines.*

LORD RAYLEIGH has pointed out, in the *Phil. Mag.* for January, 1906, that if the radiation from atoms is due to shock and recovery, the series of vibrations that would be obtained would show a simple relation between the *squares* of the frequencies, as is the case with plates and bells and other disturbed elastic vibrators, and not between the simple frequencies themselves; whereas the spectrum observations of Rydberg, and of Kayser and Runge, show that simple expressions for the *first power* of the frequency, and constant differences of frequency among a series of lines, are really applicable to the facts.

Waves of this observed kind would be emitted by electrons which radiate by reason of their unperturbed orbital motion, or other regular concomitant of the constitution of the atom, but would not be the result of oscillatory recovery from disturbance of equilibrium.

The natural frequencies of an undisturbed rotating ring of corpuscles, for instance, would be functions of the angular velocity and of the number of corpuscles in the ring, as well as of other constants.

(Thus, for instance—according to the paper mentioned near the end of last chapter—the frequencies of emission possible to two rotating corpuscles— each of mass m and charge e, rotating with angular velocity ω inside a spherical mass of positive charge Ne and radius b—will be the following six values, corresponding to their six degrees of freedom :—

$$0,\ \omega,\ c,\ c-\omega,\ c+\omega,\ \sqrt{(3c^2+\omega^2)},$$

where we have written $Ne^2/\kappa mb^3$ as c^2.

The frequencies c and ω belong to vibrations perpendicular to the plane of the orbit: the others are in that plane.

It may be noted that on the hypothesis of uniform distribution of positive electricity of density ρ throughout the sphere, the meaning of Ne/b^3 is $\frac{4}{3}\pi\rho$; and that c^2 is the ratio of this quantity to the electrochemical equivalent of an electron, in electrostatic measure. The numerical value of $c/\sqrt{N}$ is approximately 10^{16}, so that c corresponds to high ultra-violet radiation.

For three corpuscles there should be nine degrees of freedom : six in the plane, and three perpendicular thereto. The corresponding frequencies are

$$0,\ \pm\omega,\ c,\ \omega\pm c,\ \sqrt{(3c^2+\omega^2)},\ \omega\pm\sqrt{\{\tfrac{1}{2}(3c^2-\omega^2)\}}.$$

When there is no rotation, in each of the two cases considered, three of the frequencies become equal, and two vanish.)

Ordinarily, however, the constitutional radiation is excessively weak, barely perceptible ; and it is known that the radiation which emits light and produces a spectrum is the result of violence and chemical clash —that it requires something of the nature of collision to bring it out. But then anything in the nature of collision would give a series of vibrations characterised by the square of the frequency. Hence there is a difficulty.

The difficulty seems to be capable of being overcome by the suggestion—already made on page 182 —that during the chemical collision in question, the perturbation is of the nature of a disturbance of the centre of gravity of a revolving system. Such an eccentricity, or any other destruction of symmetry, would at once develop strong radiating power; but it might nevertheless leave the harmonic constituents, and peculiarities of the radiation, to be governed by the simple frequency law appropriate to the revolving constituents, rather than to the squared frequency law appropriate to their elastic recovery from vibration. In other words, the *fact* of considerable radiation would be due to the collision, but the *kind* of radiation emitted would be due to the previously existing constitution, upon which the disturbance had now conferred temporary radiating power of considerable magnitude,—temporary, because the radiation itself must speedily exhaust the energy of the disturbance and soon restore the pristine condition.

It may seem an open question whether a disturbance of centre of gravity would allow quiet revolution to continue as before, or whether it would not precipitate a catastrophe. Simple mechanical considerations show, however (*Phil. Mag.*, March, 1904, p. 264), that when the law of force is the direct distance, as it is inside the hypothetical sphere of positive electricity, no such calamity is to be expected; but that, on the contrary, a triangle or other group of corpuscles, no matter how much displaced—provided that they are not displaced so as to reach the boundary of the enclosing sphere—will continue to rotate round their common centre of gravity in the same relative position as before; while this will revolve, on its own account, round the centre of the enclosing sphere. Such a displaced group—being virtually a solitary though compound corpuscle—will now be endowed with great radiating power.

On the fairly verified hypothesis that the mass of a corpuscle is wholly electrical, it is of interest to interpret the constant c further ; it has dimensions of a frequency ; for

$$c^2 = \frac{Ne^2}{\kappa m b^3}, \quad \text{while} \quad m = \frac{2\mu e^2}{3a}.$$

So

$$c^2 = \frac{3Na}{2\mu\kappa b^3} = \frac{3Nav^2}{2b^3},$$

or

$$c = \frac{v}{b} \sqrt{\left(1\cdot 5\,N\,\frac{a}{b}\right)}.$$

Taking b as of atomic and a as of electronic dimensions, this becomes numerically

$$c = \frac{3 \times 10^{10}}{10^{-8}} \cdot \sqrt{(N \times 10^{-5})} = 10^{16} \sqrt{(N)} \text{ per second,}$$

where, since Ne measures the positive charge, N is practically the total number of corpuscles contained in the atom : not the number contained in any one ring of it.

A compound satellite will rotate round the centre of force with the same angular velocity as if it were simple,—for its mass and charge are increased in the same proportion ; there is no concentration of the charges into a single point, such as would be required to increase the mass beyond the simple multiple.

A paper by Prof. Jeans, now of Princeton, N.J., in continuation of the discussion raised by Lord Rayleigh, is to be found in the *Philosophical Magazine* for April, 1906 ; also another paper in November, 1901.

2. *Attempt to determine the number of effective corpuscles in an Atom.*

It is premature to do more than briefly refer to the remarkable attempt made by Professor J. J. Thomson, in the *Philosophical Magazine* for June, 1906, to bring forward three lines of argument which tend to show, on experimental grounds, that the number of electrons in an atom is comparable with its atomic weight, reckoning hydrogen as unity. It seems an improbable result ; but the only way to get round it is either to question the validity of the experiments and the theory applied to them, or else to realise that what is being measured is, not the total number of electrons, but the number of free or available or peculiarly constituted electrons. Cf. p. 162.

One of the arguments may be simplified as follows :

If the atom is composed of positive and negative electricities, these constituent charges will tend to be separated, against their mutual attraction, when subjected to an external electric field—such for instance as the field existing in a wave of light ; and since a light-wave is large compared with an atom, there will be time for a certain amount of this separation to be effected by each pulse. Accordingly the wave will be as it were "loaded" by the electric charges, its velocity of propagation will be reduced, and refraction will occur.

But the amount of loading, that is, the amount of effective shear or alternating separation of the two electricities, will depend on their masses; and if the mass of either the positive or the negative electricity were zero or very small, it would be shifted completely, however short the time. Accordingly a short wave would then be retarded just as much as a long one, —that is to say, there would be no discrimination of waves, and therefore no dispersion. The amount of dispersion actually experienced will therefore furnish a measure of the relative inertia of the two kinds of electricity in an atom.

The only substance to which this theory of refraction and dispersion can apply, will be one whose atoms act individually or in isolated manner; that is to say, they must be gases, the more perfect the better. Now the dispersive power of hydrogen has been measured, and is not zero; consequently there must be *some* mass both in the positive and in the negative electricity which hypothetically constitute an atom of hydrogen, and the measured amount of dispersion will enable us roughly to say what the smaller mass may be. The result is, that if M is the aggregate mass of the carriers of positive electricity, and nm the mass of the carriers of negative electricity, n being their number—so that the mass of the whole atom is $M + mn$—we get, with the aid of Ketteler's measurements for the refractive index of hydrogen for light of different wave-lengths,

$$\frac{M}{M + nm} \cdot \frac{1}{n} = 1 \text{ approximately.}$$

This shows that for a hydrogen atom n is approximately 1 (and it has to be a whole number); and it also shows that M is not small compared with nm: in fact that it is much bigger,—which is an unexpected and puzzling result.

The other lines of experimental argument seem to be confirmatory. The second one deduces, from the energy of the Röntgen radiation which is scattered by gases, as measured by Barkla of Liverpool, that molecules of air each contain approximately 25 or 28 corpuscles; which again corresponds to the molecular weight of the chief ingredient.

The third argument is based on the absorption of beta rays by metals; wherefrom it is deduced that the number of corpuscles liable to be encountered in each atom is nearly equal to the conventional atomic weight of the metal on the hydrogen scale.

This remarkable paper is the most serious blow yet dealt at the electric theory of matter, at least in its simpler and cruder form; but modes of getting round it are fore-shadowed in Chaps. XV. and XVII.

CHAPTER XXI.

VALIDITY OF OLD VIEWS OF ELECTRICAL PHENOMENA.

Now that the doctrine of electricity (at least of negative electricity) as located in small charges or charged bodies is definitely accepted, and now that a current can be treated as the locomotion of actual electricity, it may seem as if some doubt were thrown upon the doctrine, which a little time ago was spoken of as a "modern view," that the energy of an electric current resides in the space round a conductor. There is no inconsistency, however. The whole of the fields of an electron are outside itself; it is in its fields that its energy resides, and it is in the space round it that energy is conveyed when it moves; for the ether in that space is subject to the co-existence of an electric and a magnetic field. So, also, its inertia resides in space round it, for it is accounted for by the reaction experienced when its magnetic field changes,—that is, when its motion is accelerated.

In dealing with the inertia of matter it is commonly supposed that the inertia resides in the matter itself: whereas electrical inertia is known to reside in the space round the nucleus. Yet we have been emphasising and supporting the view that material inertia and electrical inertia are essentially one and the same.

Is there no inconsistency here?

The appearance of inconsistency vanishes when we come to calculate and realise how extremely local and concentrated the intense part of the field of an electron is. There is a sense in which it can be said that a moving body, for instance a vortex ring, disturbs the whole atmosphere; but any perceptible disturbance resides very near the ring. So it is with an electron. The magnetic field falls off inversely as the square of the distance from the moving nucleus; hence at a distance far less than a micro-millimetre, less even than the size of an atom, it is quite inappreciable. The whole magnetic field on which its inertia depends lies practically very close to the electron itself: it is just its extremely small size that enables this concentration to be possible, and even in a closely packed mercury atom there is practically no encroachment of the field of one electron on its neighbour's. They are all independent, each with its own inertia, almost isolated from the others : for if it were not so, the mass of a body in close chemical combination would not continue constant, but would diminish. Whether it does diminish, in the least degree, is a question perhaps worthy of attack.* Minute effects in this direction have been announced, by Heydweiler and by Landolt; but the results are doubtful.

The momentum of a moving charge at ordinary speeds is simply inversely as the radius of the sphere which holds it, as stated in Chap. II., but the localisation of this momentum, which is the point we are now considering, may be realised approximately as follows:—

The momentum depends on the co-existence and product of the electric and magnetic fields. Each

* Cf. Rayleigh, British Association Belfast, 1902.

field varies inversely as the square of the distance from the moving charge ; and their vector product is, as regards direction, perpendicular to the radius vector at any point. It is proportional, at ordinary speeds, to the sine of the angle between the radius vector and the direction of motion ; while in magnitude it falls off as the inverse fourth power of the distance. All this can be realised by common sense with very little trouble.

So, then, take a moving electron, and consider the distribution of its momentum in the space round it. Between its surface and a space of a hundred times its diameter, 99 per cent. of its momentum is contained ; because, to reckon it, we should have to integrate the factor—

$$\int_a^r \frac{4\pi r^2 dr}{r^4}.$$

But a hundred times the diameter of an electron is only 10^{-11} centimetre, that is to say, the thousandth part of the diameter of an atom. So, within the boundary of an atom, which is a hundred-thousand times an electron's diameter, there is practically none of its momentum not included.

And even in one of the comparatively closely packed atoms, e.g. in a platinum or mercury atom, the overlapping of momentum for each constituent is extremely small, since their average space apart is some thousand times the size of each constituent electron.

Consequently the assertions that an electric current is a transfer of electrons, and that the energy of a current travels in the space surrounding the moving electricity, are statements not inconsistent with each other. Nor are the statements inconsistent that the mass of a body resides in its atoms, and that inertia or momentum is a property due to the self-inductive influence of the electromagnetic field surrounding a moving electric nucleus.

The same sort of thing may be said of the way in which a current is propelled. The pace of progression of electrons through a solid may be considerable, see next section, but it is very far below the pace at which a telegraphic signal travels along a wire. They must be propelled by a lateral action, transmitted through the ether with the speed of light appropriate to the surrounding insulator, by some arrangement which "Modern Views" symbolised in the form of cog-wheels : they cannot be impelled by end thrust. The electric current is a more material entity, or has a more nearly material aspect, than was thought probable a little while since ; but all that was taught about its mode of propulsion, and the diffusion of the propelling force from outside to inside, through successive layers, as it were, of the wire—all that was taught about the paths by which the energy travels and arrives at point after point of the conductor, there to be dissipated as heat,—remains true.

Number of Ions in Conductors.

The immense number of electrons that are necessary to make up the mass of a piece of platinum, or of a lump of matter like the earth, can readily be estimated ; so, also, it is easy to imagine that an enormous number must be travelling in order to give customary strengths of current such as can readily pass through a liquid.

Through a gas, a limit is soon found to the available number ; and accordingly the conductivity of an ionised gas falls off if we call upon it to carry more than a certain current, called the saturation current. See investigations by Townsend and others briefly referred to in Chapter VII. But I am not aware of any experimental indication of such a limit in solids or liquids, at present.

In solids the pace of travel is unknown, though it has been ingeniously surmised, and is thought to be very great ; considerations of centrifugal force would make the speed of each electron during an atomic encounter equal to $e/\sqrt{(Kmr)}$ or about 10^8 centimetres per second ; views based on Maxwell's theorem about equal distribution of energy among the particles of mixed gases suggest 10^7 for the average speed of electrons at ordinary temperatures in a solid where they are free,—that is, a hundred kilometres or sixty miles per second; though since each particle is subject to constant changes of direction, this is by no means the pace of straightforward *progression*. But in liquids they are attached to atoms, and the pace of progression is known both theoretically and experimentally with considerable accuracy, and is comparable to an inch an hour for customary gradients of potential.

The total current is neu ; and to give a unit c.g.s. current at so low a speed we can reckon how many ions there must be. For $e = 10^{-20}$ electromagnetic units ; so if we take $u = 10^{-3}$ centimetre per second, the number of ions engaged in conveying the c.g.s. unit of 10 amperes is $n = 10^{23}$. But, after all, that is nothing very great; it is only about the number of atoms in a cubic centimetre of liquid. By applying a greater gradient of potential the ions can be made to move faster. By gradually narrowing down the section of a liquid conductor under a given

gradient of potential, it might be possible to get
evidence of an approach to a saturation-current-
density in liquids. The observed accuracy of Ohm's
law* under such conditions, however, is against this
experimental possibility.

Conclusion.

The subject is very far from exhausted, but I
must not attempt to cover more ground. The
most exciting part of the whole is the explanation of
matter in terms of electricity, the view that electricity
is, after all, the fundamental substance, and that what
we have been accustomed to regard as an indivisible
atom of matter is built up out of it; that all atoms—
atoms of all sorts of substances—are built up of the
same thing. In fact the theoretical and proximate
achievement of what philosophers have always sought
after, viz., a *unification of matter* is offering itself
to physical enquiry. But it must be remembered
that although this solution is strongly suggested
it is not yet a completed proof. Much more
work remains to be done before we are certain that
mass is due to electric nuclei. If it is, then
we encounter another surprising and suggestive
result, namely that the spaces inside an atom are
enormous compared with the size of the electrical
nuclei themselves which compose it; so that an atom
can be regarded as a complicated kind of astronomical
system,—like Saturn's ring, or perhaps more like a
nebula; with no sun, but with a large number of equal
bodies possessing inertia and subject to mutual
electric attractive and repulsive forces of great magnitude, to replace gravitation. The radiation of a

* FitzGerald and Trouton, *Brit. Assoc. Reports*, 1886, 1887, 1888.

nebula may be due to shocks and collisions somewhat like the X-radiation from some atoms.

The disproportion between the size of an atom and the size of an electron is vastly greater than that between the sun and the earth. If an electron is depicted as a speck one-hundredth of an inch in diameter, like one of the full-stops on this page for instance, the space available for the few hundred or thousand of such constituent dots, to disport themselves inside an atom, is comparable to a hundred-feet cube; in other words, an atom on the same scale would be represented by a church 160 feet long, 80 feet broad, and 40 feet high,—in which therefore the dots would be almost lost. And yet on the electric theory of matter they are all of the atom that there is; they " occupy " its volume in the sense of keeping other things out, as soldiers occupy a country; they are energetic and forceful though not bulky; and in their mutual relations they constitute what we call the atom of matter; they give it its inertia; they enable it to cling on to others which come within short range, with the force we call cohesion; and by excess or defect of one or more constituents they exhibit chemical properties and attach themselves with vigour to others in like or rather opposite case.

That such a hypothetical atom, composed only of sparse dots, can move through the ether without resistance is not surprising. They have links of attachment with each other, but, so long as the speed is steady, they have no links of attachment with the ether; if they disturb it at all, in steady motion, it is probably only by the simplest irrotational class of disturbance which permits of no detection by any optical means.* Nor do

See Lodge, *Phil. Trans.* 1893, vol. 184, pp. 750-754; also vol. 189, p. 166.

they tend to drag it about. All known lines of mechanical force reach from atom to atom, they never terminate in ether; except indeed at an advancing wave front. At a wave front is to be found one constituent of a mechanical pressure of radiation whose other constituent acts on the source. This is an interesting but essentially non-statical case, and it leads away from our subject.

As to the nature of an electron, regarded as an ethereal phenomenon, it is too early to express any opinion. At present it is not clear why a positive charge should cling so tenaciously in a mass, while an outstanding negative electron should readily escape and travel free. Nor is the nature of gravitation yet understood. When the electron theory is complete, to the second order, or some higher *even* order, of small quantities—it is complete now to the second order if electrons may be treated as geometrical points,—it is hoped that the gravitative property also will fall into line and form part of the theory; at present it is an empirical fact which we observe without understanding; as has been our predicament not only since the days of Newton but for centuries before : though we did not, before Newton, know its importance in the cosmic scheme.

Attention has hitherto been chiefly concentrated on the freely moving active negative ingredient,—the more sluggish positive charges are at first of less interest,—but the behaviour of electrons cannot be fully and properly understood without a knowledge of the nature and properties of the positive constituent too. According to Larmor, positive charge must be the mirror-image of negative charge, in essential constitution.

The positive electron has not, so far as I know, been

as yet observed free. Some think it cannot exist in a free state, that it is in fact the rest of the atom of matter from which a negative unit charge has been removed ; or, to put it crudely—that "electricity" repels "electricity," and "matter" repels "matter," but that Electricity and Matter in combination form a neutral substance which is the atom of matter as we know it. Such a statement is an extraordinary and striking return to the views expressed by that great genius, Benjamin Franklin. On any hypothesis those views of his are of exceeding interest, and show once more the kind of prophetic insight which we have had occasion to notice in discoverers before (Appendix H and Chap. IV.). Undoubtedly we are at the present time nearer to the view of Benjamin Franklin than men have been at any intervening period between his time and ours.

The view that an atom is composed of an equal number of interleaved or inter-revolving positive and negative electrons—that view is not Franklin's ; nor is it as yet anything but a guess. To make it more, work must be done upon the nature and properties of the positive charge ; and the positive electron, if it exists, must be dragged experimentally to light.

Especially must the inner ethereal meaning both of positive and negative charges be explained : whether on the notion of a right-and-left-handed self-locked intrinsic wrench-strain in a Kelvin gyrostatically-stable ether, elaborated by Larmor,* or on some hitherto unimagined plan. And this will entail a quantity of exploring mathematical work of the highest order.

* See *Ether and Matter*, p. 326 ; or *Phil. Trans.* 1894, pp. 810, 811, and 1897, pp. 209-212.

APPENDIXES.

APPENDIX A.

Calculation of the Inertia of an Electric Charge.

Let a spherical conductor of radius a, carrying a charge of electricity, move forward with moderate speed u; meaning by moderate speed anything distinctly less than the speed of light: it constitutes a current element of magnitude eu, and its circuit is closed by displacement currents in the surrounding dielectric. For electric lines of force arise in the medium in front, and subside in the medium behind, and so a displacement of electricity takes place from fore to aft, to compensate the motion forward; and the lines of displacement are identical with the magnetic lines due to a short magnet. A charge may be said to travel carrying its electrostatic lines with it, or it may be said to be constantly generating an electrostatic field in front and destroying one behind.

When an electric field thus moves, partly laterally, it generates a magnetic field—in the present instance in circular lines round the line of motion;—for the moving charge is an element of a linear current. The generation of these magnetic lines acts so as to oppose the current which produced them; though so long as they continue steady they exert no effect on it. When they subside, however, they tend to prolong the current which maintained them. Consequently, if the moving charge (or current) tries to stop, its retardation meets with obstruction; it is constrained to persist by the subsidence of the magnetic field which its motion excited and maintains. Its velocity is not resisted, there is nothing equivalent to friction, but its acceleration + or − is obstructed, an effect precisely analogous to inertia. If it is at rest it will need force to start it, and if it is in motion its motion will persist, even against force, for a time.

The charge acts, therefore, as if it had inertia, and we can proceed to calculate its amount.

While moving it is a current, and will be surrounded by rings of magnetic force, whose intensity, at any point with polar co-ordinates r, θ, referred to the line of motion as axis and the moving charge as origin, will be the quite ordinary expression (with eu for the current-element instead of Cds)—

$$H = \frac{eu \sin \theta}{r^2}.$$

The ordinary expression for the electrostatic force at the same point is

$$E = \frac{e}{\kappa r^2},$$

(see note at end of Chapter I. with reference to the insertion of κ) and if the motion is slow this value will be preserved; but if it is rapid the electric field gets weaker along the axis and stronger equatorially, having been shown by Mr. Heaviside (*Philosophical Magazine*, April, 1889) to be given by the following expression—

$$E = \frac{e}{\kappa r^2} \cdot \frac{1 - (u/v)^2}{\{1 - (u \sin \theta / v)^2\}^{\frac{3}{2}}},$$

where v is the velocity of light.

The strength of the magnetic field will be similarly modified in this case; but the simplest mode of stating that is to express it in terms of E, and to say that *always*—

$$H = \kappa E u \sin \theta.$$

The rate of transmission of energy will be the vector product of E and H; and the whole magnetic energy,—that is the whole kinetic energy due to the current, *i.e.*, due to the motion—will be obtained by integrating the ordinary expression $\mu H^2/8\pi$, all over space outside the charged sphere, viz., from a to ∞ all round. Its charge is assumed to be superficial, so that no energy is inside. In the general case this expression is a little long, but in the most important case, when the speed of motion u is decidedly less than the speed of light v, it is quite simple, and the working may as well be given:—

Kinetic energy

$$= \int_a^\infty \frac{\mu H^2}{8\pi} d(\text{vol.}) = \frac{\mu e^2 u^2}{8\pi} \int_0^\pi \int_0^{2\pi} \int_a^\infty \frac{\sin^2\theta}{r^4} dr . r d\theta . r \sin\theta d\phi$$

$$= \frac{\mu e^2 u^2}{8} \int_0^{2\pi} \int_a^\infty \frac{\cos^2\theta - 1}{r^2} dr . d\cos\theta = \frac{\mu e^2 u^2}{3a}.$$

Comparing this with mechanical kinetic energy, $\tfrac{1}{2}mu^2$, we see that the charge on the sphere confers upon it additional kinetic energy, as if its mass were increased on account of the charge by the amount—

$$m = \frac{2\mu e^2}{3a}.$$

This may also be written—

$$m = \frac{2}{3}\frac{\mu\kappa . e^2}{\kappa a} = \frac{2}{3v^2} . e . \frac{e}{\kappa a} = \frac{2}{3v^2} \times \text{charge} \times \text{potential},$$

or, $\frac{3}{4} mv^2 =$ the electrostatic energy of the charge :

supposed a spherical shell of electricity.

In other words, the mass equivalent to the charge is such that if it were a piece of matter with constant inertia travelling at the speed of light, its kinetic energy would be half as great again as the potential energy of the electric charge when standing still.

APPENDIX B.

The Electric Field due to a Moving Magnet.

If a short bar magnet or uniformly magnetised sphere (its moment M being the intensity of magnetisation × the volume of the sphere) moves along axially—that is in the direction of its magnetisation—with velocity u, it generates circular lines of electric force all centred upon its axis, much as a moving charge generates circular lines of magnetic force. If there is a conducting path round any such circle, then the motion of a magnet along its axis will generate a current in it; but if there be no conductor, the motion will only result in an electric displacement which subsides when the magnet stops.

The intensity of the magnet's field at any point along its axis is well known to be $2M/r^3$; at any point on its equatorial plane it is $-M/r^3$; and in any intermediate direction it is, as regards magnitude alone—

$$H = \frac{M}{r^3} \sqrt{(1 + 3\cos^2\theta)}.$$

All this holds for the moving as for the stationary magnet, provided its speed does not approach that of light.

The electric force at the same point is—

$$E = \frac{3}{2}\frac{Mu}{r^3}\sin 2\theta$$

$$= 3Hu\ \frac{\sin\theta}{\sqrt{(4 + \tan^2\theta)}}.$$

The electrostatic energy resulting will be the integral of $\kappa E^2/8\pi$ everywhere outside the moving magnetised sphere of radius a, viz.—

$$\text{Energy} = \frac{\kappa}{8\pi}\iiint\left(\frac{3\mathrm{M}u}{r^3}\right)^2 \sin^2\theta\cos^2\theta\, dr\,.\,rd\,\theta\,.\,r\sin\theta\, d\phi$$

$$= \frac{\kappa \mathrm{M}^2 u^2}{5a^3} = \frac{\mathrm{M}^2}{5\mu a^3}\left(\frac{u}{v}\right)^2.$$

The displacement acts like an elastic strain set up in the dielectric, storing the above energy statically; and so long as the magnet continues moving steadily the electric displacement exerts no force upon it. But acceleration will be resisted; for if the magnet begins to go faster it sets up more displacement, and the act of setting this up constitutes a transient current, which opposes the motion as long as the acceleration continues, but dies out the instant the motion becomes steady again.

Conversely if the motion of the magnet began to slacken, the electric strain would begin to subside, and its subsidence would constitute an inverse transient current which would assist the motion, *i.e.*, oppose the slackening. In other words, the variations of the circular electric strain in the surrounding medium confer upon a moving magnet a spurious or apparent momentum, in addition to its real mechanical momentum ; and thus the elastic strain itself may be said to represent a spurious or apparent inertia due to magnetisation, in addition to any real mechanical inertia which the body holding the magnetism may itself possess. And the amount of this extra inertia is—

$$m = \frac{2\kappa \mathrm{M}^2}{5a^3} = \frac{2}{5}\frac{\mathrm{M}^2}{\mu a^3 v^2} = \frac{8}{15}\cdot\frac{\pi\mathrm{I}\mathrm{M}}{\mu v^2}$$

$$= \frac{2}{5}\frac{\mathrm{H}_0\mathrm{M}}{v^2},$$

where I is the intensity of magnetisation, and H_0 the intensity of the field, inside the substance of a uniformly magnetised sphere of radius a and magnetic moment M.

The equivalent mass moving with the velocity of light would therefore have an energy equal to two-fifths of the intrinsic energy of the magnetised sphere.

APPENDIX C.

On Electricity and Gravitation and Dimensions.

Referring back to an article of mine in the *Philosophical Magazine* for November, 1882, page 358, we find the fundamental and necessary relation between constants stated thus, where M shall stand for magnetic pole and γ for Cavendish's gravitation constant—

$$e^2/\kappa \equiv M^2/\mu \equiv \gamma m^2 = Fl^2,$$

F being force and l being length.

If it is now going to turn out that a mass is composed of electric charges, it might seem as if e and m were quantities of the same nature, and were only numerically connected; whence it would follow that κ and γ were of similar kind. In other words, Faraday's dielectric constant would become closely related to Cavendish's gravitation constant, and *weight* as well as *mass* would be traced to electricity; but such a deduction is unwarranted, there is nothing to prevent essentially different properties of the ether being involved in the two kinds of force—gravitative and electric.

As to the nature of the gravitation constant itself, we have—

$$\gamma = \frac{Fl^2}{m^2} = \frac{l^3}{mt^2} = \frac{v^2}{m/l} = \frac{\text{sq. of velocity}}{\text{linear density}} = \frac{\text{energy/mass}}{\text{mass/length}}.$$

It is clear that if gravitation is in any sense of electric origin it must be a second order disturbance, or some higher *even* order, superposed upon the main electric effect, and be independent of sign. It would, in fact, depend upon e^4. For the gravitative force between two electrons at distance r would be—

$$F_1 = \gamma \frac{m^2}{r^2} = \frac{\gamma}{r^2} \left(\frac{2\mu e^2}{3a} \right)^2.$$

The electric force between the same two electrons at the same distance is—

$$F_2 = \frac{e^2}{\kappa r^2}.$$

Therefore the ratio of the gravitative to the electric force at any distance is constant and equal to—

$$\frac{F_1}{F_2} = \frac{4\mu^2 \kappa \gamma}{9a^2} e^2 = \frac{4\mu \gamma e^2}{9a^2 v^2} = \frac{2\gamma}{v^4} \cdot F_0,$$

where F_0 is the electric force between two spherical electrons in contact, and v is the velocity of light.

Numerically this ratio of the two forces is—

$$\frac{F_1}{F_2} = \kappa \gamma \left(\frac{m}{e} \right)^2 = \frac{1}{9 \times 10^{20} \times 1\cdot5 \times 10^7} \left(\frac{1}{10^7} \right)^2 = 10^{-42},$$

so the electric force exceeds the gravitative as much as the globe of the earth exceeds in bulk an ultra-microscopic object.

When there is an agglomeration of electrons of opposite sign, their electric influence at a distance disappears, but their gravitative potential will be proportional to the sum of the squares of the charges. So with 10^{21} mixed electrons in each of two bodies, at any distance apart, the gravitative force between them will equal the electric force between two single electrons at the same distance.

In my 1885 Report to the British Association on Electrolysis, page 745, the following statement is made:—If the opposite electricities were extracted from a milligramme of water and given to two spheres one mile apart, those two spheres would attract each other with a force equal to the weight of 12 tons.

APPENDIX D.

Dimensions of e/m Ratio.

The reciprocal of the electrochemical equivalent of a substance, e/m, may be expressed as regards dimensions in several ways, one of which exhibits it as a certain large numerical multiple of $\sqrt{(\kappa\gamma)}$, the geometric mean between Faraday's dielectric constant and Cavendish's gravitation constant. For hydrogen, this numerical multiple is of the order 10^{18}; for silver 10^{16}.

Another way is obtained by writing—

$$e^2 = \kappa F l^2 = \frac{ml}{\mu},$$

whence it follows that—

$$\frac{m}{e} = \frac{\mu e}{l} = \sqrt{\left(\frac{\mu m}{l}\right)},$$

and so e/m can be expressed in $\sqrt{\left(\dfrac{\text{centimetres}}{\mu \text{ grammes}}\right)}$.

The artificiality of these dimensions is due to the fact that e and m have been conventionally measured in different ways; m is measured by ratio of applied external force to acceleration, while e is measured by repulsive force self-exerted on a similar charge at given distance.

If we express μ as a density (see "Modern Views of Electricity," Appendix p), the electrochemical equivalent comes out as expressible in grammes per square centimetre, that is to say a surface density.

It is noteworthy that while $\sqrt{(\kappa\mu)}$ is of the same dimensions as $1/v$, $\sqrt{(\kappa\gamma)}$ corresponds to $1/\epsilon$, where ϵ is an electrochemical equivalent.

APPENDIX E.

Electric Saturation, etc.

In my Report on Electrolysis to the British Association for 1885 (see the Aberdeen volume, pp. 762, 763), I call attention to the possibility that an atomic theory of electricity would give rise to a maximum charge possible on a given area. The maximum surface-density would be attained when every atom was polarised so that its atomic charge faced outwards; and for a solid or liquid it would be very great. For the charge on each being 10^{-10}, and the number of atoms per square centimetre being 10^{16}, it follows that the maximum surface density possible is $\sigma = 10^6$ electrostatic units per square centimetre. The corresponding gradient of potential would be $4\pi\sigma = 10^7$, or 3,000 megavolts per centimetre; and the corresponding tension would be $2\pi\sigma^2 = 6 \times 10^{12}$ c.g.s. $= 40,000$ tons to the square inch. Of course no dielectric would stand this pressure, but absolute vacuum might.

In practice, therefore, it follows that when a surface is charged highly, only an exceedingly small percentage of the molecules are polarised with their charges facing outwards. For instance, common air breaks down when the tension rises to a value $2\pi\sigma^2 = \frac{1}{2}$ gramme per square centimetre $= 400$ c.g.s.; wherefore the maximum σ in ordinary air is 8 electrostatic units per square centimetre; and this quantity would be afforded by the facing outwards of 10^{11} molecules, or one in every hundred thousand of a solid surface, or about a tenth per cent. of those in air.

It is shown on p. 760 of my 1885 B.A. Report on Electrolysis, that a potential gradient of the order 1 volt over molecular distance is sufficient to overcome atomic attraction and effect decomposition in liquids. Any liquid which is a conductor throws the whole applied stress on to a molecular layer contiguous to an electrode, and accordingly something of the order of a volt or two difference of potential between electrodes in such a liquid is required, and is sufficient, for decomposition.

APPENDIX F.

Size of Orbit of Radiating Electron.

Consider two electrons of opposite sign revolving round each other with luminous frequency n at any distance r; or better, consider a free negative electron revolving round a comparatively fixed equal positive charge attached to an atom, at distance r.

The force between them is $e^2/\kappa r^2$, so the acceleration is—

$$\frac{e^2}{\kappa r^2} \cdot \frac{3a}{2\mu e^2} = \frac{3av^2}{2r^2}.$$

But the acceleration is also expressible as $4\pi^2 n^2 r$. Therefore—

$$r^3 = \frac{3av^2}{8\pi^2 n^2} = \frac{3a\lambda^2}{8\pi^2} = \frac{3 \times 10^{-13}}{80}(6 \times 10^{-5})^2 = 10^{-23},$$

which is "Kepler's third law" for the case, and indicates that the distance at which luminous frequency is attainable is the atomic distance 10^{-8} centimetre; in other words, that the electron is roaming over the surface of the atom. If it got nearer to the centre of force than this, without penetrating any of the attracting substance, it would have to revolve quicker; and such rapid oscillations may be excited among the internal paired electrons by shocks and collisions, or other perturbation.

The most important aspect of the above calculation is that it corresponds with the hypothesis that the whole of the mass of an electron is electric, and none of it material or unexplained; for it shows that a pure electron is able to revolve at distances of the molecular order with luminous frequency.* The square of the wave length emitted is proportional to the cube of the radius vector; provided the plane of the orbit contains the centre of force,—otherwise there may be constrained motion of smaller amplitude, analogous to that of a conical pendulum.

APPENDIX G.

The Radiating Power of a Steadily Revolving Electron.

Consider an electron revolving as above (Appendix F) in an orbit of atomic dimensions b with luminous frequency $n = \omega/2\pi$; and calculate its radiating power.

By considering separately the electric and magnetic forces due to such a particle, at any point of space, and then applying Poynting's theorem as to the convection of energy wherever the two fields coexist, we get, as the rate of transmission of energy past a point whose polar coordinates, referred to centre and axis of orbit, are r, θ, the mean value

$$\frac{\mu}{v} \cdot \frac{1 + \cos^2\theta}{8\pi} \cdot \frac{e^2 b^2 \omega^4}{r^2}.$$

And integrating this all over the sphere of radius r we get, as the total emission of energy per second—

$$\frac{2\mu e^2 b^2 \omega^4}{3v},$$

* See Lodge in *The Electrician* for March 12th, 1897, vol. 38, p. 644.

which may be written,—with a as the radius, m the mass, and $\ddot{u}$ as the acceleration, of the revolving electron—

$$\frac{ma\ddot{u}^2}{v},$$

where v is the velocity of light.

The fundamental expression for the amount of energy emitted per second as waves in the ether, by a moving charge e, was given by Larmor in *Phil. Mag.*, December, 1897, page 512, so far as I know for the first time; also in *Æther and Matter*, page 227, namely—

$$\frac{2\mu e^2}{3v}\, \ddot{u}^2.$$

This agrees with the above calculation, since $\ddot{u} = u^2/b = b\omega^2$; $\ddot{u}$ being acceleration. Now μe^2 may be taken as 10^{-40} gramme-centimetre, according to most recent measurements; and in a circular orbit of radius r the acceleration is

$$\ddot{u} = (2\pi n)^2\, b = 40(5 \times 10^{14})^2 10^{-8} = 10^{23} \text{ c.g.s.;}$$

therefore the radiating power of a single electron, so moving, is

$$\frac{2\mu e^2}{3v}\, \ddot{u}^2 = \frac{2 \times 10^{-40}}{9 \times 10^{10}} \times 10^{46} = 2 \times 10^{-5} \text{ ergs per second.}$$

But the total available energy possessed by the revolving electron of linear dimensions a is only

$$\frac{\mu e^2}{3a}\, u^2 = \frac{\mu e^2}{3a}(2\pi n b)^2,$$

namely its kinetic energy (for of course it cannot radiate away or dissipate its electrostatic energy), and this amounts to—

$$\frac{10^{-40}}{3 \times 10^{-13}}(2\pi \times 5 \times 10^{14} \times 10^{-8})^2 = 3 \times 10^{-13} \text{ ergs,}$$

its velocity being 3×10^7 centimetres per second, or one-thousandth that of light. So if the electron were isolated from any supply of energy, and if it could maintain the pace, it would at this rate radiate away all its kinetic energy in 10^{-8} of a second, that is to say in three or four million revolutions. This may seem a rapid rate of cooling, but it is not surprising for an isolated and luminous atom: it is a Hertzian vibrator or emitter of simple type. The number of revolutions which an electron must make in one second, in order to emit sodium light, is about 8000 times the number of seconds which have passed since the Christian era.

We may express the ratio of the radiating power of a single electron to its total kinetic energy, by the fraction—

$$\frac{2a}{v}\left(\frac{\dot{u}}{u}\right)^2 = \frac{2a}{v}(2\pi n)^2 = 8\pi^2 n \frac{a}{\lambda} = 70 \text{ million per second.}$$

In any large assemblage of atoms the radiation is not free and unrestrained, nor is it unmaintained, like this; but it must always be considerable at anything like luminous frequency, and it is proportional to the fourth power of the frequency. At a frequency which emits a wave ten times as long as a luminous wave, the radiating power of a revolving electron is only one ten-thousandth of that above calculated, but even so it is very significant; so there must be compensation of some kind or a substance could not permanently exist. The criterion that a molecule shall not be destroyed by radiation-losses is given in the concluding sentence of Larmor's paper above quoted : *Phil. Mag.*, Dec., 1897.

The subject of radiation from a symmetrical group of electrons was pursued above in Chapter XIX.

The radiating power of an electron suddenly stopped by a collision is of course much greater than the above, and is estimated in Chapter IX. To get copious Rontgen rays.

the stopping distance must be comparable with the electrons' own diameter; which accordingly accounts for the extreme thinness and consequent penetrating character of the emitted pulse.

APPENDIX H.

Faraday's Prophetic Nomenclature.

Students of the life of Faraday will remember that when he discovered the rotation of the plane of polarisation by a magnetic field applied to dense bodies in which light travelled along the lines of force,—wresting the secret from nature by strong and pertinacious experimental research that would not be denied, though the time was as yet by no means ripe for comprehension of the fact when it was discovered—he labelled his discovery in a fit of enthusiasm, "The Magnetisation of Light and the Illumination of Magnetic Lines of Force": a label which puzzled contemporaries for a long time.

It is difficult to see what meaning he can have attached to these phrases; and for many years afterwards they appeared unsuitable misnomers, indicating a foggy conception of his own discovery.

It is not likely that his state of mind was really at all clear on the subject, and probably he would at a later stage have been willing to plead guilty to a less than lucid mode of conceiving the phenomenon; which nevertheless always specially pleased him, though when it was reduced to a mere rotation of the plane of polarisation, it seemed to many mathematicians and physicists to have lost its unique and surprising interest. It must always be remembered, however, that interest was never lost by either Lord Kelvin or Clerk Maxwell, and that it was the chief fact which incited Maxwell, many years later, to begin developing his electro-magnetic theory of light.

But how do the titles strike us now? Do they not indicate some extraordinary unconscious insight, such as is frequently experienced by a great discoverer in the enthusiasm of discovery? Remember that the Hall effect, the Zeeman effect, the Aurora Borealis, and Faraday's rotation are all closely connected with each other—by means of the electron theory.

In the cathode ray tube the flying electrons are deflected by a cross magnetic field; or if they fly along the lines they are twisted into a spiral path round them. In the Aurora Borealis this effect is carried out in the upper region of the air on a gigantic scale, and the earth's magnetic "lines of force are illuminated" by flying electrons from the sun entangled and guided by them. In the Hall effect this same influence is felt by the slowly moving crowd of electrons as they are handed on from one atom to the next, causing a curvature of the current path—in which either positive or negative may predominate. In the Zeeman effect the same cause operates on the revolving and vibrating electrons, associated with a radiating atom and constituting a source of light; wherefore we may truly say that the "light is magnetised," for the source of light is magnetised directly, and the effect is impressed on and retained by the light emitted, and is made visible by spectrum analysis.

The first intimation of that magnetic influence on light which lies at the base of all these at first sight apparently diverse phenomena was detected by Faraday in his slight differential rotation of the plane of polarisation in one direction or the other by a magnet, according as the positive or the negative element in the dense substance was most affected.

Hence the title which he affixed to his discovery—"The

illumination of the lines of magnetic force and the magnetisation of light"—may be regarded as a prophetic flash of
genius.

A not altogether dissimilar flash has already been
referred to, when Crookes hinted prematurely that in the
cathode rays we had something like corpuscular light, and
also like matter in a fourth state, neither solid, liquid, nor
gaseous. For, whether quite right or not, he was far
more right than the critics of those days who presumed
to deride him.

APPENDIX J.

On the β-rays from Radium.

Magnetic deflexion of Beta-rays.

The following diagram illustrates the spreading out
into a sort of spectrum of the Beta-rays or electric
particles shot out with different velocities from radium.
Some are only slightly bent,—being very speedy.

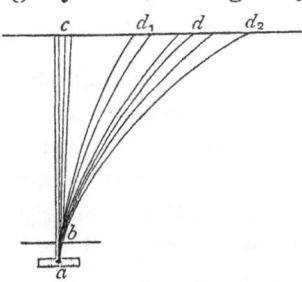

a represents the source of radiation ;

b an aperture through which the rays have to pass;

c the impression which would be produced on the
photographic plate with a magnetic field absent, and

d_1, *d*, d_2 the impress on the same plate due to the
deflected rays.

In a uniform magnetic field the rays will all be segments of circles, and it is easy to estimate the radius of curvature of the ray corresponding to any point in the spectrum, since only one circle can be drawn through three given fixed points (Euclid, IV. 5 or III. 25), such points for instance as a, b, and d; so, these three points being known, the circle of which the ray forms a part is known, and its radius of curvature is therefore determined.

Electric charge carried by Beta-rays.

The fact that ordinary cathode-rays are negatively electrified was proved most conclusively by Perrin with an apparatus diagrammatically represented in the figure; where a represents the cathode,

    b an earthed diaphragm with small aperture, and

    c a Faraday cavity or hollow vessel, arranged to catch the rays in its interior and convey the charge to an electroscope.

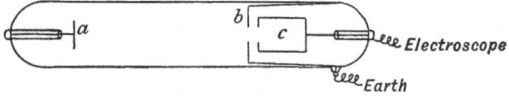

It is not so easy to prove that the beta rays emitted by radium are electrified, because the air and everything in the neighbourhood is rendered conducting by their impact, but Professor Curie succeeded in establishing the fact by enclosing a piece of metal completely in solid paraffin, and showing that when exposed to the rays this metal became charged.

Strutt exhibits the converse effect in a simple and interesting manner by hanging up a radium tube in a vacuum and attaching to it a pair of gold leaves. As the electric rays are shot away by the radium the leaves diverge with an opposite charge; they go on diverging to the full extent, and collapsing when they touch the boundary, for as long as the radium lasts, which must be many centuries.

Strutt's apparatus was shown in figure 22, page 177.

For a good and clear elementary account of the phenomena exhibited by radium, especially of all those most important facts which have reference to things other than electrons,—such as alpha-rays, the emanations, and the transmutations of matter,—the book on *The Becquerel Rays and the Properties of Radium*, by the Hon. R. J. Strutt, may satisfactorily be referred to, as an introduction to Prof. Rutherford's treatise.

APPENDIX K.

Note on the Behaviour of a Charge Moving Nearly at the Speed of Light.

According to investigations by Larmor in the *Phil. Trans.*, 1897, pp. 228-9, and also according to the investigations of Mr. Searle (*Phil. Mag.*, Oct. 1897) a charge does not re-distribute itself on a moving body when its speed becomes great, but the lines of force bend or are deflected towards the equator, without remaining normal to the surface whence they start. Any uncertainty on this head seems to have been due to a natural confusion between the electric force acting on the *convected* charge and the etherial force which would act on charges at rest, or would cause corresponding 'displacement' in force ether; the former must be normal to a perfect conductor, but the latter need not: an assertion in which we may trace some analogy to the fact that in a moving medium rays of light are not perpendicular to their wave fronts.

There is no question but that the lines of force bend back towards the equator, as stated by me in 1902, but I assumed that this deflexion of the lines would entail their moving up nearer to the equator of the sphere, so as to leave the poles bare of charge, in order that the lines might still continue radial. I admit that the lines of force need not continue radial close to the sphere; but, in so far as the sphere changes its shape, there should still be some unimportant redistribution of the charge as the speed increases. Mr. Searle calculates that whereas a sphere at rest acts as if its charge were at a central point, this equivalent point opens out into a uniformly charged line, forming a medial and small portion of its diameter, when the sphere is in motion; as the velocity increases, the length of this line gradually increases also, until the speed equals that of light, when it fits the sphere exactly. But this leaves out of account a distortional change in the sphere itself, to which I will presently refer.

The fact is that the behaviour of a charged body moving at enormous speed may be treated exactly in the manner of elementary potential theory for a charged ellipsoid at rest.

It is doubtful whether the term "inertia" remains useful under these conditions: it is perhaps best to reserve it for the ordinary case when mass is constant; for, as Mr. Searle points out, three different estimates of inertia can be made:—one the ratio of force to acceleration, another the ratio of momentum to velocity, and a third as the ratio of kinetic energy to half the square of velocity. In ordinary matter, as is well known, and for slow electric motions, these three estimates are one and the same; but for violent electric motions they become different; though it should be realised how small the difference is, until the speed of light is very closely approached; so that in no material case of great velocity or great acceleration that has ever been practically dealt with—as, for instance, the case of a cannon-ball stopped by armour plate—is any sort of unusual effect to be expected ; even on the hypothesis that matter is entirely electrically composed. Nevertheless, now that among free corpuscles in a vacuum tube, or among those expelled from radium, it is becoming practically possible to attain these high speeds, and even to begin to base crucial determinations upon them, it becomes necessary to consider the matter more carefully.   In a publication at Göttingen in January, 1902, Dr. Abraham has thus discriminated what he calls "longitudinal" from what he calls "transverse" inertia ; making inertia depend not only on the speed but on the direction of acceleration ; each direction having a different inertia of its own.

And all these results are still further complicated by a consideration of the effect of acceleration itself, which, whenever it is violent, gives rise to some perceptible radiation, involving dissipation of energy ; and this radiation loss of energy, though it will be primarily represented in the motion as a resistance or velocity term, may secondarily have an effect on inertia ;—probably, however, quite a small and subordinate effect in all practical cases, and no effect at all so long as motion occurs with uniform speed in a straight line : for then there is no radiation. But then, of course, under those conditions it is not possible to test or measure the inertia of a body ; it is only when the motion is either curved or changed in some way that inertia becomes prominent, and then there is necessarily some, though usually very small, radiation too.

When magnetic deflexion of a charged body is being observed at ultra-high speeds it may be asked whether it is possible for the ordinary expression for the force exerted on a current by a moving field to be departed from.

The ordinary expression for deflecting force is euH at low speeds, for a charge e moving at speed u across a magnetic field of intensity H; but whether this simple expression is departed from at high speeds must be a question of etherial dynamics: the procedure of Larmor, based on the principle of 'Least Action' (see *Æther and Matter*, p. 97), would give an answer in the negative,— which agrees with the assumption of Lorentz. It has been suggested by others that for speeds at which $(u/v)^2$ becomes sensible, we must use the more complex expression for deflecting force:—

$$eH\frac{v^2-u^2}{u}\left(\frac{v}{2u}\log\frac{v+u}{v-u}-1\right).$$

This, however, at low speeds reduces not to the usual simple value, but to one-third of that value, viz. $\frac{1}{3}Heu$; and Professor Schuster in the *Philosophical Magazine* for January, 1897, calls attention to the variety of numerical estimates of this quantity given by different varieties of the main theory. But it appears to be now considered that there is no real ambiguity and that Larmor's view is correct.

As has been said above, at high speeds, not only does effective inertia vary with speed, but it has different values for different *directions* of acceleration relative to the line of motion; the value of what Abraham calls "transverse inertia," which expresses reaction to a transverse deflecting force, is quoted by Kaufmann in *Comptes Rendus*, vol. cxxxv, p. 577, writing it with m_0 as the equivalent inertia for slow motion, and with β as the ratio u/v—the ratio of the velocity of the particles to the velocity of light—thus

$$m=\frac{3m_0}{4\beta^2}\left(\frac{1+\beta^2}{2\beta}\log\frac{1+\beta}{1-\beta}-1\right),$$

and is the formula appropriate to his experiments. All this is in fact deducible at once by the usual Lagrangian dynamical method from Mr. Heaviside's expression— *Electrical Papers*, vol. ii., p. 514—for kinetic energy; viz. an expression equivalent to $\frac{1}{2}u^2$ multiplied by the following quantity:—

$$\frac{\mu e^2}{a} \cdot \frac{1-r}{4r}\left(1+\frac{2r-\frac{1}{2}}{1-r}+\frac{(2r-\frac{1}{2})\tan^{-1}\sqrt{\left(\frac{r}{1-r}\right)}}{\sqrt{\{r(1-r)^3\}}}\right),$$

r being the squared speed ratio u^2/v^2.

In Larmor's original treatment of electrical inertia, *Phil. Trans.* 185A (Aug. 1894), pp. 806-818, there was no reason whatever for anticipating velocities greater than one-tenth of light, so a simple inertia theory seemed amply sufficient at that time; the mass being a permanent constant associated with the electron and dependent on its structure.

It is the hope of seeing somewhat into *structure* that has made the recent experiments on its modification at high speeds so interesting.

Professor J. J. Thomson's corresponding formula for momentum is quoted above in the text, page 133, and simplified; and in simplified form it may be re-quoted:

It amounts to this, that the mass of an electric charge e on a small non-conducting sphere of radius a, moving with a speed $u = v\sin\theta$, is—

$$m = \frac{\mu e^2}{4a(1-\cos 2\theta)}\left\{(1-2\cos 2\theta)\frac{2\theta}{\sin 2\theta}+(2-\cos 2\theta)\right\},$$

which we have tabulated on page 145 (see also page 133) in the form—

$$m = \frac{2\mu e^2}{3a} \cdot \phi(\theta),$$

226 DISTORTION [APP. K

and to a certain extent it must be considered experimentally verified by Kaufmann's results. The function $\phi(\theta)$ has the value unity when $u=0$, or even when u/v is small.

APPENDIX L.

Distortion due to High-speed Motion through the Ether.

Mr. Searle, in the *Philosophical Magazine*, October, 1897, points out that the simplest charged body when in motion is not a sphere, but an oblate spheroid, oblate in the direction of motion, with its axes in the ratio $\sqrt{\left(1-\dfrac{u^2}{v^2}\right)}, 1, 1$; and that this produces on all points outside itself exactly the same effect as a point charge at its centre: wherefore such a spheroid in motion at the speed u takes the place of the sphere in electrostatics. He calls this a Heaviside ellipsoid, because Mr. Heaviside first indicated its importance in the theory of moving charges.

But it is well known that a spheroid of this kind is exactly what a sphere in rapid motion would automatically become, on the FitzGerald-Lorentz theory; viz., that hypothesis which was started in order to account for the negative result in Michelson's experiment, by postulating a change of dimensions in solid bodies according to their direction of motion through the ether. This hypothesis, shown to be plausible by Lorentz, became a definite theory when Larmor proved (*Æther and Matter*, ch. xi.) that on the electric theory of matter—that is, assuming that the whole inertia of matter was electric—not only was such a change of dimensions reasonably likely, as FitzGerald had perceived, and likely also to be of the right amount to give a compensating effect, and precisely zero resultant, in the Michelson experiment, but that the change was a necessary consequence of dynamical molecular theory.

The change of dimensions, thus imagined and justified, is gradually coming to be accepted as certainly true; and it is interesting to note that a sphere in motion, by reason of being subject to this amount of distortion, still retains its property of being the simplest geometrical body, so far as the distribution of its electric field is concerned. True, it is then no longer a sphere; but no measuring instrument could possibly show its distortion, because all standards of measurement would share it. It is a remarkable thing that this imperceptible and unmeasurable uniform distortion of all matter should ever have been discovered: nothing but an ethereal process could have dragged it to light. Nevertheless dragged to light it has been, by the combined testimony of electrical theory and of optical experiment.

APPENDIX M.

Constitution of Electrons.

In continuation of the subject treated of in appendices K and L we may consult Larmor, *Phil. Trans.* 190A (April, 1897), pp. 225-8. In *Æther and Matter* (1898), ch. xi., he substituted an improved (dynamical) investigation, applicable to a system of molecules in the most complicated motion, wherein he claims to cover all possible cases, on the single hypothesis that the electrons in an atom are at distances apart compared with which the diameters of their structure are very small, so that they may be treated as points. On that hypothesis uniform convection of a system at any speed however high should show no internal influence, *provided* the constitution of matter is *wholly* electric, *i.e.* provided atoms are active only through the æther. The very important (because purely electrical) experiment of Trouton and Noble is a case in confirmation. So is the absence of influence on either of the two phenomena, magnetic rotation and double refraction (Rayleigh and Brace).

It is true that this way of explaining the absence of any second order aberration-effect was received with scepticism by Poincaré (Paris Congress, 1900), whose criticism was that if in future the third order effect also needed annulment some new artificial arrangement could still be added on, to do even that. The result of the argument has now, however, gained the independent support of Lorentz (1904), whose investigation has suggested, what in fact is easily verified, that Larmor's work proves to be exact: his restriction to the second order being unnecessary. He was hardly concerned, however, to go further, since the hypothesis of the infinitesimal effective size of the electron already limited the investigation,—nor does it seem that Lorentz's work really carries the matter further. Larmor's attitude has been all along (*Phil. Mag.*, June, 1904), that this provisional hypothesis holds the field until some effect of uniform convection presents itself: every new negative result is a steady corroboration of it: but like any other physical representation it must ultimately reach the limits of its application.

Only negative electrons are known in the free state. It remains unsettled whether this is due to some one-sidedness in the experimental means hitherto employed, or whether physical nature is in reality intrinsically unsymmetrical. If the positive electron were a region of

singularity (beknottedness) of dimension large compared
with the negative, it would be but a feeble agent in the
transformation of energy, and thus could readily escape
detection. Cf. *Æther and Matter*, § 122. But the
absolute structure of an electron is probably very unlike
the distributions of electric charge on spheres and ellip-
soids that have hitherto been taken to represent it.

The theory of a simple molecule appeared for the first
time in *Phil. Trans.*, 13th August, 1894, pp. 806-818;
inertia purely electric, on p. 807; estimate of size and
dimensions, and electrodynamics of orbital motions, on
p. 814; Hall effect, etc., on p. 815. There does not
appear to be anything in Lorentz's papers, either of 1892
or 1895, to correspond to the ideas there introduced.

With reference to a theory of the Zeeman effect: the
simplest type of illustrative system, amenable to calculation, would be a statical one, in which the electrons
would all vibrate round positions of stable equilibrium.
But this requires extraneous supporting force, if they are
in empty space. A definite subdivision of the periods
has been worked out for J. J. Thomson's statical illustration, in which this innate instability of a statical system
of negative electrons (Earnshaw's theorem) is obviated
by supposing them inside a region of continuous positive
electrification,—which may be taken to represent the positive electron. If the volume density of this is small, the
resulting stability will be but slight, and a small displacement will upset the system and lead to its break-up;—
which is unlikely, perhaps unreasonable.

But here, as before, it is only isotropic configurations,
in the same sense as in the hydrokinetics of solids in
fluids,—namely, those with which an associated quadratic
function must be wholly isotropic, including the configurations of the regular solids,—that split the lines definitely

230   CONSTITUTION OF ELECTRONS   [APP. M.

instead of merely broadening them: other types will only do so when they are all similarly orientated to the inducing magnetic field.

The experiments of Kaufmann held out a hope that we should get to know something definite about intrinsic electron structure. They have proved sufficiently the fundamental fact that a free electron has no independent material sub-stratum: they have made it very probable that it may be provisionally represented as a region of electricity spherically stratified: also that convection of an electron does not affect the volume relations of this distribution (Bucherer); but unfortunately it has not proved possible to push on the precision of the experiment so as to get information beyond this.

We must still be content to treat the electron as a point, or at most as a spherical electric aggregate of some sort whose volume does not undergo shrinkage. But the indirect evidence afforded by the entire absence of convection effects of many kinds, such as the earth's motion might be supposed to excite, is strongly in favour of the provisional theory that the ultimate elements of which atoms are constituted are—in their dynamical relations— purely æthereal structures of some sort which probably we cannot yet adequately imagine.

GLASGOW: PRINTED AT THE UNIVERSITY PRESS BY ROBERT MACLEHOSE AND CO. LTD.